# Soil Health Series: Volume 1 Approaches to Soil Health Analysis

# Soil Health Series: Volume 1 Approaches to Soil Health Analysis

*Edited by Douglas L. Karlen, Diane E. Stott, and Maysoon M. Mikha*

Soil Science Society of America

WILEY

*Editorial Correspondence:*
Soil Science Society of America, Inc.
5585 Guilford Road, Madison, WI 53711-58011, USA
soils.org

*Registered Offices:*
John Wiley & Sons, Inc., 111 River Street, Hoboken, NJ 07030, USA

For details of our global editorial offices, customer services, and more information about Wiley products, visit us at www.wiley.com.

Wiley also publishes its books in a variety of electronic formats and by print-on-demand. Some content that appears in standard print versions of this book may not be available in other formats.

*Library of Congress Cataloging-in-Publication Data applied for*
Paperback: 9780891189800
doi: 10.2136/soilhealth.vol1

Cover Design: Wiley
Cover Image: © Joshua Miller, DPH, PhD, Steve Culman, Timothy Clough, Pedro Ferrari Machado, Negar Tafti

Set in 9.5/12.5pt STIXTwoText by Straive, Pondicherry, India

## Dedication

These books are dedicated to Dr. John W. Doran, a retired USDA-ARS (Agricultural Research Service) Research Soil Scientist whose profound insight provided international inspiration to strive to understand the capacity of our fragile soil resources to function within ecosystem boundaries, sustain biological productivity, maintain environmental quality, and promote plant and animal health.

Understanding and quantifying soil health is a journey for everyone. Even for John, who early in his career believed soil quality was too abstract to be defined or measured. He initially thought soil quality was simply too dependent on numerous, uncontrollable factors, including land use decisions, ecosystem or environmental interactions, soil and plant management practices, and political or socioeconomic priorities. In the 1990s, John pivoted, stating he now recognized and encouraged the global soil science community to move forward, even though perceptions of what constitutes a *good* soil vary widely depending on individual priorities with respect to soil function. Continuing, he stated that to manage and maintain our soils in an acceptable state for future generations, *soil quality* (*soil health*) must be defined, and the definition must be broad enough to encompass the many facets of soil function.

John had profound impact on our careers and many others around the World. Through his patient, personal guidance he challenged everyone to examine soil biological, chemical, and physical properties, processes, and interactions to understand and quantify soil health. For Diane, this included crop residue and soil enzyme investigations, and for Maysoon, interactions between soil physical and biological processes mediated by water-filled pore space. Recognizing my knowledge of soil testing and plant analysis on Midwestern soils, as well as root-limiting, eluviated horizons and soil compaction in Southeastern U.S. soils, John encouraged me to develop a strategy to evaluate and combine the biological, chemical, and physical indicators that have become pillars for soil quality/health assessment. The Soil Management Assessment Framework (SMAF) was the first generation outcome of this challenge.

Throughout his life, John endeavored to involve all Earth's people, no matter their material wealth or status, in translating their lifestyles to practices that strengthen social equity and care for the earth we call home. Through development of the "soil quality test kit" John fostered transformation of soil quality into *soil health* by taking his science to farmers, ranchers, and other land managers. These two volumes have been prepared with that audience in mind to reflect the progress made during the past 25 years. Special thanks are also extended to John's life mate Janet, daughter Karin, son-in-law Michael, grandchildren Drew and Fayth, and all of his friends for their encouragement, patience and support as he continues his search for the "holy grail" of soil health. Without John's inspiration and dedication, who knows if science and concern for our fragile soil resources would have evolved as it has.

Thank you, John – you are an inspiration to all of us!

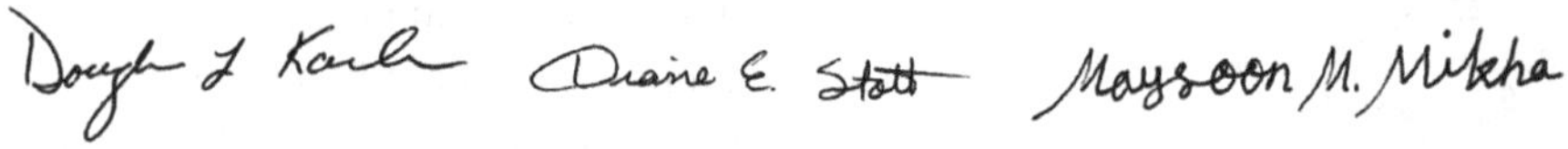

# Contents

# Foreword

Soil science receives increasing attention by the international policy arena and publication of this comprehensive "Soil Health" book by the Soil Science Society of America (SSSA) and Wiley International is therefore most welcome at this point in time. Striving for consensus on methods to assess soil health is important in positioning soil science in a societal and political discourse that, currently, only a few other scientific disciplines are deeply engaged in. Specifically, increasing the focus on sustainable development provides a suitable "point on the horizon" that provides a much needed focus for a wide range of activities. Sustainable development has long been a likeable, but still rather abstract concept. The United Nations General Assembly acceptance of seventeen Sustainable Development Goals (SDGs) by 193 Governments in 2015 changed the status of sustainable development by not only specifying the goals but also defining targets, indicators, and seeking commitments to reach those goals by 2030 (https://www.un.org/sustainabledevelopment-goals). In Europe, the Green Deal, accepted in 2019, has targets and indicators corresponding to those of the SDGs (https://ec.europa.eu/info/strategy/european-green-dealsoil).

So far, soil scientists have not been actively engaged in defining SDG targets, which is unfortunate considering soil functions contribute significantly to ecosystem services that, in turn, contribute to the SDGs. The connections are all too obvious for soil scientists, but not necessarily so for scientists in other disciplines, politicians, or the public at large. For example, adequate production of food (SDG2) is impossible without healthy soil. Ground- and surface-water quality (SDG6) are strongly influenced by the purifying and infiltrative capacities of soils. Carbon capture through increases in soil organic carbon (SOC) is a major mechanism contributing to the mitigation of an increasingly variable climate (SDG13) and living soils as an integral part of living landscapes are a dominant source of biodiversity (SDG15) (Bouma, 2014; Bouma et al., 2019). With complete certainty, we can show that healthy soils make better and more effective contributions to ecosystem services than unhealthy ones! This also applies when considering the recently introduced Soil Security concept, which articulates the 5 C's: soil capability, condition, capital, connectivity, and codification (Field et al., 2017). A given soil condition can be expressed in terms of soil health, whereas soil capability defines potential conditions, to be achieved by innovative soil management, thus increasing soil health to a characteristically attainable level for that particular soil. Healthy soils are a capital asset for land users; connectivity emphasizes interactions among land users, citizens, and politicians that are obviously important, especially when advocating measures to increase soil health that may initially lack societal support. Finally, codification is important because future land use rules and regulations could benefit by being based on quantitative soil health criteria, thus allowing a reproducible comparison between different soils.

These volumes provide an inspiring source of information to further evaluate the soil health concept, derive quantitative procedures that will allow more effective interaction among land users, and information needed to introduce soil science into laws and regulations. The introductory chapters of Volume 1 present a lucid and highly informative overview of the evolution of the soil health movement. Other chapters discuss data needs and show that modern monitoring and sensing techniques can result in a paradigm shift by removing the traditional data barriers. Specifically, these new methods can provide large amounts of data at relatively low cost. The valuable observation is made that systems focusing only on topsoils cannot adequately represent soil behavior in space and time. Subsoil properties, expressed in soil classification, have significant and very important effects on many soil functions. Numerous physical, chemical and biological methods are reviewed in Volume 2. Six chapters deal with soil biological methods, correctly reflecting the need to move beyond the traditional emphasis on physical and chemical assessment methods. After all, soils are very much alive!

The book *Soil Health* nicely illustrates the "roots" of the soil health concept within the soil science profession. It also indicates the way soil health can provide "wings" to the profession as a creative and innovative partner in future environmental research and innovation.

**Johan Bouma**
**Emmeritus Professor of Soil Science**
**Wageningen University**
**The Netherlands**

## References

Bouma, J. (2014). Soil science contributions towards Sustainable Development Goals and their implementation: Linking soil functions with ecosystem services. *J. Plant Nutr. Soil Sci.* 177(2), 111–120. doi:10.1002/jpln.201300646

Bouma, J., Montanarella, L., and Vanylo, G.E. (2019). The challenge for the soil science community to contribute to the implementation of the UN Sustainable Development Goals. *Soil Use Manage.* 35(4), 538–546. doi:10.1111/sum.12518

Field, D.J., Morgan, C.L.S., and Mc Bratney, A.C., editors. (2017). *Global soil security*. Progress in Soil Science. Springer Int. Publ., Switzerland. doi:10.1007/978-3-319-43394-3

# Preface

This two-volume series on Soil Health was written and edited during a very unique time in global history. Initiated in 2017, it was intended to simply be an update for the "Blue" and "Green" soil quality books entitled *Defining Soil Quality for a Sustainable Environment* and *Methods for Assessing Soil Quality* that were published by the Soil Science Society of America (SSSA) in the 1990s. In reality, the project was completed in 2020 as the United States and world were reeling from the Covid-19 coronavirus pandemic, wide-spread protest against discriminatory racial violence, and partisan differences between people concerned about economic recovery versus protecting public health.

Many factors have contributed to the global evolution of soil health as a focal point for protecting, improving, and sustaining the fragile soil resources that are so important for all of humanity. Building for decades on soil conservation principles and the guidance given by Hugh Hammond Bennett and many other leaders associated with those efforts, soil health gradually is becoming recognized by many different segments of global society. Aligned closely with soil security, improving soil health as a whole will greatly help the United Nations (UN) achieve their Sustainable Development Goals (SDGs). Consistent with soil health goals, the SDGs emphasize the significance of soil resources for food production, water availability, climate mitigation, and biodiversity (Bouma, 2019).

The paradox of completing this project during a period of social, economic, and anti-science conflicts associated with global differences in response to Covid-19, is that the pandemic's impact on economic security and life as many have known it throughout the 20th and early 21st centuries is not unique. Many of the same contentious arguments could easily be focused on humankind's decisions regarding how to use and care for our finite and fragile soil resources. Soil conservation leaders such as Hugh Hammond Bennett (1881–1960), "Founder of Soil Conservation," W. E. (Bill) Larson (1921–2013) who often stated that soil is "the thin layer covering the planet that stands between us and starvation," and many current conservationists can attest that conflict regarding how to best use soil resources is ancient. Several soil science textbooks, casual reading books, and other sustainability writings refer to the Biblical link between soil and humankind, specifically that the very name "Adam" is derived from a Hebrew noun of feminine gender (*adama*) meaning earth or soil (Hillel, 1991). Furthermore, Xenophon, a Greek historian (430–355 BCE) has been credited with recording the value of green-manure crops, while Cato (234–149 BCE) has been recognized for recommending the use of legumes, manure, and crop rotations, albeit with intensive cultivation to enhance productivity. At around 45 CE, Columella recommended using turnips (perhaps tillage radishes?) to improve soils (Donahue et al., 1971). He also suggested land drainage, application of ash (potash), marl (limestone), and planting of clover and alfalfa (N fixation) as ways to make soils more productive. But then, after Rome was conquered, scientific agriculture, the arts, and other forms of culture were stymied.

Advancing around 1500 yr, science was again introduced into agriculture through Joannes Baptista Van Helmont's (1577–1644 CE) experiment with a willow tree. Although the initial data were misinterpreted, Justice von Liebig (1803–1873 CE) eventually clarified that carbon (C) in the form of carbon dioxide ($CO_2$) came from the atmosphere, hydrogen and oxygen from air and water, and other essential minerals to support plant growth and development from the soil. Knowledge of soil development, mineralogy, chemistry, physics, biology, and biochemistry as well as the impact of soil management (tillage, fertilization, amendments, etc.) and cropping practices (rotations, genetics, varietal development, etc.) evolved steadily throughout the past 150 yr. **SO**, what does this history have to do with these 21st Century Soil Health books?

First, in contrast to the millennia throughout which humankind has been forewarned regarding the fragility of our soil resources, the concept of soil health (used interchangeably with soil quality) per se, was introduced only 50 yr ago (Alexander, 1971). This does not discount outstanding research and technological developments in soil science such as the physics of infiltration, drainage, and water retention; chemistry of nutrient cycling and availability of essential plant nutrients, or the biology of N fixation, weed and pest control. The current emphasis on soil health in no way implies a lack of respect or underestimation of the impact that historical soil science research and technology had and have for solving problems such as soil erosion, runoff, productivity, nutrient leaching, eutrophication, or sedimentation. Nor, does it discount contributions toward understanding and quantifying soil tilth, soil condition, soil security, or even sustainable development. All of those science-based accomplishments have been and are equally important strategies designed and pursued to protect and preserve our fragile and finite soil resources. Rather, soil health, defined as an integrative term reflecting the "capacity of a soil to function, within land use and ecosystem boundaries, to sustain biological productivity, maintain environmental quality, and promote plant animal, and human health" (Doran and Parkin, 1994), is another attempt to forewarn humanity that our soil resources must be protected and cared for to ensure our very survival. Still in its infancy, soil health research and our understanding of the intricacies of how soils function to perform numerous, and at times conflicting goals, will undoubtedly undergo further refinement and clarification for many decades.

Second, just like the Blue and Green books published just twenty years after the soil health concept was introduced, these volumes, written after two more decades of research, continue to reflect a "work in progress." Change within the soil science profession has never been simple as indicated by Hartemink and Anderson (2020) in their summary reflecting 100 yr of soil science in the United States. They stated that in 1908, the American Society of Agronomy (ASA) established a committee on soil classification and mapping, but it took 6 yr before the first report was issued, and on doing so, the committee disbanded because there was no consensus among members. From that perspective, progress toward understanding and using soil health principles to protect and preserve our fragile soil resources is indeed progressing. With utmost gratitude and respect we thank the authors, reviewers, and especially, the often-forgotten technical support personnel who are striving to continue the advancement of soil science. By developing practices to implement sometimes theoretical ideas or what may appear to be impossible actions, we thank and fully acknowledge all ongoing efforts. As the next generation of soil scientists, it will be through your rigorous, science-based work that even greater advances in soil health will be accomplished.

Third, my co-authors and I recognize and acknowledge soil health assessment is not an exact science, but there are a few principles that are non-negotiable. First, to qualify as a meaningful, comprehensive assessment, soil biological, chemical, and physical properties and processes must all be included. Failure to do so, does not invalidate the assessment, but rather limits it to an assessment of "soil biological health", "soil physical health", "soil chemical health", or some combination thereof. Furthermore, although some redundancy may occur, at least two different indicator measurements should be used for each indicator group (*i.e.*, biological, chemical, or physical). To aid indicator selection, many statistical tools are being developed and evaluated to help identify the best combination of potential measurements for assessing each critical soil function associated with the land use for which an evaluation is being made.

There is also no question that any soil health indicator must be fundamentally sound from all biological, chemical, physical and/or biochemical analytical perspectives. Indicators must have the potential to be calibrated and provide meaningful information across many different types of soil. This requires sensitivity to not only dynamic, management-induced forces, but also inherent soil properties and processes reflecting subtle differences in sand, silt, and clay size particles derived from rocks, sediments, volcanic ash, or any other source of parent material. Soil health assessments must accurately reflect interactions among the solid mineral particles, water, air, and organic matter contained within every soil. This includes detecting subtle changes affecting runoff, infiltration, and the soil's ability to hold water through *capillarity*– to act like a sponge; to facilitate gas exchange so that with the help of $CO_2$, soil water can slowly dissolve mineral particles and release essential plant nutrients– through *chemical weathering;* to provide water and dissolved nutrients through the soil *solution* to plants, and to support exchange between oxygen from air above the surface and excess $CO_2$ from respiring roots.

Some, perhaps many, will disagree with the choice of indicators that are included in these books. Right or wrong, our collective passion is to start somewhere and strive for improvement, readily accepting and admitting our errors, and always being willing to update and change. We firmly believe that starting with something good is much better than getting bogged down seeking the prefect. This does not mean we are discounting any fundamental chemical, physical, thermodynamic, or biological property or process that may be a critical driver influencing soil health. Rather through iterative and ongoing efforts, our sole desire is to keep learning until soil health and its implications are fully understood and our assessment methods are correct. Meanwhile, never hesitate to hold our feet to the refining fire, as long as collectively we are striving to protect and enhance the unique material we call soil that truly protects humanity from starvation and other, perhaps unknown calamities, sometimes self-induced through ignorance or failing to listen to what our predecessors have told us.

*Douglas L. Karlen (Co-Editor)*

## References

Alexander, M. (1971). Agriculture's responsibility in establishing soil quality criteria In: *Environmental improvement– Agriculture's challenge in the Seventies.* Washington, DC: National Academy of Sciences. p. 66–71.

Bouma, J. (2019). Soil security in sustainable development. *Soil Systems.* 3:5. doi:10.3390/soilsystems3010005

Donahue, R. L., J. C. Shickluna, and L. S. Robertson. 1971). *Soils: An introduction to soils and plant growth.* Englewood Cliffs, N.J.: Prentice Hall, Inc.

Doran, J.W., Coleman, D.C., Bezdicek, D.F., and Stewart, B.A., editors. (1994). *Defining soil quality for a sustainable environment. Soil Science Society of America (SSSA) Special Publication No. 35.* Madison, WI: SSSA Inc.

Doran, J.W., and Parkin, T.B. (1994). Defining and assessing soil quality. In: J.W. Doran, D.C. Coleman, D.F. Bezdicek, and B.A. Stewart, editors, *Defining soil quality for a sustainable environment. SSSA Special Publication No. 35.* Madison, WI: SSSA. p. 3–21. doi:10.2136/sssaspecpub35

Doran, J.W., and Jones, A.J. (eds.). (1996). *Methods for assessing soil quality. Soil Science Society of America (SSSA) Special Publication No. 49.* Madison, WI: SSSA Inc.

Hartemink, A. E. and Anderson, S.H. (2020). 100 years of soil science society in the U.S. *CSA News* 65(6), 26–27. doi:10.1002/csann.20144

Hillel, D. (1991). *Out of the earth: Civilization and the life of the soil.* Oakland, CA: University of California Press.

# 1

# Soil Health: An Overview and Goals for These Volumes

*Douglas L. Karlen*, Diane E. Stott, Maysoon M. Mikha, and Bianca N. Moebius-Clune*

## Synopsis of Two-Volume Book

Farmers and ranchers, private sector businesses, non-governmental organizations (NGOs), academic-, state-, and federal-research projects, as well as state and federal soil conservation, water quality and other environmental programs have begun to adopt soil health as a unifying goal and promote it through workshops, books, and public awareness meetings and campaigns. The driver is an increased awareness that soil resources are crucial for not only meeting global demand for high-quality food, feed, and fiber but also to help mitigate more extreme weather events and to protect water and air quality, wildlife habitat, and biodiversity.

Volume 1 briefly reviews selected "Approaches to Soil Health Analysis" including a brief history of the concept, challenges and opportunities, meta-data and assessment, applications to forestry and urban land reclamation, and future soil health monitoring and evaluation approaches.

Volume 2 focuses on "Laboratory Methods for Soil Health Analysis" including an overview and suggested analytical approaches intended to provide meaningful, comparable data so that soil health can be used to guide restoration and protection of our global soil resources.

* Disclaimer: Mention of names or commercial products in this document does not imply recommendation or endorsement by the U.S. Department of Agriculture.

*Soil Health Series: Volume 1 Approaches to Soil Health Analysis*, First Edition.
Edited by Douglas L. Karlen, Diane E. Stott, and Maysoon M. Mikha.

## Introduction

Soil health research, books, workshops, websites, press releases, and other forms of technology transfer materials have made rural and urban producers and consumers of all ages more aware of soil resources and the services they provide. Innovative farmers and ranchers, the private sector, non-governmental organizations (NGOs), academic, state, and federal researchers, and policymakers around the world are becoming more aware of how properly functioning soils more effectively respond to: (1) changing climate patterns and more extreme weather events (Paustian et al., 2016); (2) increasing demands for abundant, high-quality food, feed, and fiber to meet needs of an increasing global population (Doran, 2002), and (3) the need to protect water, air, wildlife, plant, and microbial biodiversity (Andrén & Balandreau, 1999; Havlicek & Mitchell, 2014).

Enhancing global soil health will improve humankind's capacity to maintain or increase crop yield, achieve better yield stability, reduce purchased input costs, and enhance critical ecosystem services (Boehm & Burton, 1997). Striving for improved soil health is not only important for croplands, but also for pastures, native rangelands, orchards, and forests (Herrick et al., 2012; Chendev et al., 2015; Gelaw et al., 2015; Vitro et al., 2015). Yet, there is still a lot of confusion and uncertainty regarding soil health in the U.S. and around the world. One reason is that soils are complex and perform many different functions that respond to changes in the same properties and processes in different and sometimes conflicting ways. For example, what may be considered good soil health characteristics for crop productivity (e.g., well aggregated, porous with good water infiltration, efficient nutrient cycling) may not be optimum for water quality if high infiltration rates and/or macropores result in rapid transport of contaminants to surface or subsurface water resources. Similarly, no-tillage as a single practice may improve soil health by increasing soil organic carbon (SOC), but improper management decisions (e.g., timing, equipment size, lack of living roots) or unanticipated weather patterns (e.g., multiple freeze–thaw cycles) may increase compaction and runoff compared to using a moderate fall tillage operation. For those reasons, soil health assessment and management must always be holistic, striving to balance trade-offs, and accounting for biological, chemical, and physical property and process changes to be useful and meaningful for regenerative and sustainable soil management and protection of our fragile resources.

The concept of soil health is not new (see Figure 2.1 of Chapter 2). It has evolved from both indigenous knowledge derived over millennia through trial and error, and over a century of soil and agronomic research focused on soil management, soil conservation, soil condition, soil quality, soil tilth, soil security, and similar topics. Fundamental roots of soil health principles can be traced to the time of Plato (Hillel, 1991) and Columella, a prominent writer about agriculture within the Roman Empire (~40 to 60 BCE). Current soil health efforts reflect the enormous efforts given by people such as Hugh Hammond Bennett, founder of the Soil Conservation Service (SCS) now known as the Natural Resources Conservation Service (NRCS). Soil health activities can be traced to soil conservation efforts implemented in response to the Dust Bowl and other natural events. As a result, it has become a mantra to focus people's attention on the soil beneath their feet (Carter et al., 1997; Montgomery, 2007). Unfortunately, as acknowledged 25 yr ago (Doran and Jones, 1996), soil health was and continues (Chapter 3) to be a controversial topic.

Many current soil health activities began to emerge in the 1970s (Alexander, 1971). In part, they were accelerated by the 1973 U.S. oil embargo which increased energy and nitrogen (N) fertilizer prices (Warkentin & Fletcher, 1977). Escalating N fertilizer prices led to renewed interest among soil and agronomic researchers regarding how the soil microbial community might be enhanced to help supply crop-available N rather than continuing to depend on costly fertilizer inputs (Gregorich & Carter, 1997; Tilman, 1998). The Food Security Act of 1985 also introduced new incentives to encourage producers to implement minimum- or no-tillage conservation practices to reduce soil erosion, thus increasing farmer and society focus on the importance of soils for producing the food and fiber humans need and. For maintaining the ecosystems on which all life ultimately depends (National Research Council, 1993).

In contrast to soil quality efforts during the 1990s and early 2000s, a major driver of soil health projects from 2011 to 2020 has been investment by private industry. This can be partially explained by the rapid increase in corporate social responsibility reporting between 2011 and 2020 (Sustainability Reports, 2019). Consumer demand and sustainable, responsible shareholder investment pressures have driven this increase in reporting—which has created a corporate need for transparency in the environmental impact from agricultural production systems.

Increased public awareness of soil health has opened avenues to productive partnerships between industry, governmental, grower and conservation organizations due to the ability to create win-win-win scenarios between farm economic, environmental improvement (e.g., water quality, greenhouse gas emissions, biodiversity) and social outcomes (e.g., AgSolver and EFC Systems development of 'Profit Zone Manager' and its incorporation into the FieldAlytics platform for field data management; ANTARES– Enabling Sustainable Landscape Design project linking soil health and the continual improvement of sustainable operating bioenergy supply systems).

A leader in building public-private-partnerships focused on soil was the Soil Renaissance which was initiated to reawaken public interest and awareness of the importance of soil health in vibrant, profitable and sustainable natural resource systems. Founded as a Farm Foundation and Noble Research Institute collaboration, it sought to make maintenance and improvement of soil health (https://www.farmfoundation.org/projects/the-soil-renaissance-knowledge-to-sustain-earths-most-valuable-asset-1873-d1/) the cornerstone of land use management. The Soil Health Partnership (SHP) (https://www.soilhealthpartnership.org/science/) initiated by the National Corn Growers Association (NCGA), Walton Family Foundation, Monsanto (Bayer), Environmental Defense Fund (EDF) and the Nature Conservancy (TNC) in 2014 was another leader. Soil Renaissance endeavors have been carried on through the formation of the Soil Health Institute which has provided leadership for a North American project to evaluate soil health measurements (Norris et al., 2020). Meanwhile, the SHP has focused on using science and data to work directly with farmers to adopt practical agricultural practices including (i) cover crops, (ii) conservation tillage, and (iii) advanced nutrient management to improve the economic and environmental sustainability of the farm. Administered by the NCGA, the partnership has more than 220 working farms enrolled in 15 states and one Canadian province. Collectively SHP, SHI, and other regional, state and local partnerships have created an exponential increase in recognition and adoption of soil and crop management practices that can protect, improve, and sustain our fragile soil, water, and air resources.

Many additional soil health projects, partnerships, and investment opportunities have arisen across the United States (e.g., The Wells Fargo Innovation Incubator, or IN$^2$, The Soil Coalition initiated by Rabobank, a.s.r. and Vitens, and S2G Ventures). The IN$^2$, a technology incubator and platform co-administered by the U.S. Department of Energy's National Renewable Energy Laboratory (NREL), was initiated with six startups focused on agriculture technology solutions, while S2G's portfolio companies are on a mission to better align the food system to meet changing consumer demands. Collectively, these partnerships and projects have sent farmer and consumer market demand signals across the entire agricultural supply chain. Subsequently, soil health products and services have followed the market demand signals. For example: General Mills now brands products with information regarding soil health and carbon sequestration (General Mills, 2020); BASF began focusing on soil health when they launched Poncho Votivo 2.0 a treatment designed to protect corn seeds and increase microbial activity in the soil (BASF, 2020); and Nutrien Ltd, an agricultural retail company that distributes potash, nitrogen, and phosphate products worldwide for agricultural, industrial, and feed customers. Nutrien which serves the agriculture industry worldwide, purchased Waypoint Analytical, Inc.—a soil science company—in 2018 to expand soil health analyses for farmers (Nutrien Ltd., 2018). These investments as well as those by the Environmental Defense Fund (EDF), Midwest Row Crop Collaborative (MWRCC), National Wheat Foundation, Foundation for Food and Agriculture Research (FFAR), Natural Resources Conservation Service (NRCS), Minnesota Corn Growers Association, and Iowa Corn Growers Association at the regional, state and local level have created partnerships supporting an exponential increase in recognition and adoption of soil and crop management practices that can protect, improve, and sustain our fragile soil, water, and air resources.

Historically, a significant soil health development during the 1980s and 1990s was the Canadian publication entitled "The Health of Our Soil" (Acton & Gregorich, 1995) which was one of the first broad-scale, organized efforts to provide land managers information on implementing SH-improving practices. Following those Canadian efforts, several U.S. soil scientists developed a definition of soil quality and recommended assessment methods to characterize how tillage and other crop management decisions were affecting soil resources (e.g., Doran et al., 1994; Doran & Jones, 1996; Karlen et al., 1997). The importance of soil biology was recognized as integral to improving the understanding and measurement of soil quality, but optimum methods to assess soil microbial communities were still being developed (Pankhurst et al., 1997). As the capacity to quantify soil biology indicators improved, discussions of SQ were replaced by the term soil health which was used to communicate to both producers and consumers the importance of understanding and managing soil as a living ecosystem. Consistent with that messaging, the NRCS ultimately defined soil health as "the continued capacity of the soil to function as a vital living ecosystem that supports plants, animals, and humans" (USDA-NRCS, 2019a).

The purpose and scope for this two-volume series (I. Approaches to Soil Health Analysis and II. Laboratory Methods for Soil Health Assessment) are to review advancements in soil health since Defining Soil Quality for a Sustainable Environment (Doran et al., 1994) and Methods for Assessing Soil Quality (Doran & Jones, 1996) were published 25 yr ago. Our goal for Volume 1 is to provide agricultural and conservation communities an update that will help identify appropriate soil health indicators for various soil processes important for agriculture, forest, and reclamation functions. Volume 2 provides standardized, science-based guidelines for sampling and procedures for assessing soil organic carbon (SOC), aggregate stability and compaction, pH and salinity, nutrient availability, as well as microbial processes, diversity, and community structure. Numerous scientific publications and technical outreach activities have contributed to the evolution of soil health and are cited in the various chapters. Four relatively recent examples are Basche and DeLonge (2017) who focus on soil hydrologic effects of continuous living covers, Congreves et al. (2015) who reported on long-term impacts of tillage and crop rotations on soil health, McDaniel et al. (2014) who used a meta-analysis to examine crop diversity effects on soil microbial biomass and soil organic matter (SOM) dynamics, and Turmel et al. (2015) who quantified crop residue management effects on soil health. Collectively, the information in those publications and numerous others can and will be used to produce consistent meaningful guidelines that can be understood and used by producers to improve their long-term soil and crop management practices. This two-volume series is also intended to help producers and land managers more fully understand their soil's response to human management. This is essential to move beyond current, broadly available soil-testing methods that generally focus only on chemical extractions to assess nutrient status and make nutrient management recommendations.

## Why is Soil Health Important?

Investing in regenerating, improving, or sustaining soil health will result in a broad array of benefits for producers and the public. Those benefits include: carbon sequestration and potential mitigation of and adaptation to climate change; increased soil organic carbon (SOC) stocks; increased water infiltration, storage, and availability to plants; reduced runoff, water-induced soil erosion, and flooding; more efficient nutrient cycling and pest suppression; reduced need for agricultural inputs; protection of groundwater, surface water, and air resources, including reduced dust storm events; increased biodiversity and resilience; long-term economic viability; and perhaps most importantly, food security, defined as sustained, reliable productivity needed to provide the food, feed, fiber, and fuel resources for an increasing world population (Glæsner et al., 2014; DeLong et al., 2015; Lal, 2015).

Aggressively pursuing continued advancement of publicly available soil health testing is critical because current chemical-based soil-testing approaches do not provide a complete view of the soil physical, chemical, and biological interactions and constraints that influence overall soil function. Fortunately, over the last four decades, laboratory methods have been developed and refined for studying, quantifying, and monitoring biological and physical indicators. This makes it possible to use a combination of field observations and laboratory tests to identify factors affecting a variety of soil, water, air, and plant resource concerns.

Many new biological and physical soil health assessment methods are still being refined and validated by the research community, but several indicators and laboratory methods are gradually becoming available through agricultural soil testing laboratories for assessing how well soil processes are functioning. Technical Note 450–03 (USDA-NRCS, 2019b) was published by NRCS to provide recommendations reflecting current best available methods compiled through meetings and working groups involving more than 100 scientists who collaborated in multiorganizational workshops co-organized by USDA-NRCS Soil Health Division and members of the Soil Renaissance effort. The Technical Note laboratory methods are used as part of the Conservation Innovation Grants Soil Health Demonstration Trials minimum dataset and in a newly available Soil Testing Conservation Activity (CA 216) standard (USDA-NRCS, 2020c). Standardization of methods enables nationwide baseline measurements to be obtained that can be used for monitoring, and to guide soil and crop management when combined with a soil health assessment framework that provides soil and climate adapted interpretations of raw laboratory values. These book volumes are intended to summarize current best available methods and to identify gaps in our current understanding of soil health measurement and assessment. They are also intended to facilitate and increase in public awareness of soil health assessment and to provide insight and science-based methods for those evaluations. Furthermore, by encouraging soil health assessments, we hope these volumes will ultimately result in compilation of a national dataset that can be used to support multiple public and private soil health goals and document the value of public and private investments in such assessments.

While qualitative or semi-quantitative field observations can be used for preliminary identification of soil health constraints or to improve soil and crop management practices, identifying specific underlying causes and/or the management practices needed to address them, often requires quantitative laboratory analysis. We anticipate information in these volumes will be used by a wide group of stakeholders including producers, consultants, technical service providers, conservation planners, and other private and public agricultural service providers, conservation groups, researchers, industry, policymakers, and the general public. Uses will include: (1) identifying soil health problems and planning and implementing soil health management systems; (2) innovating, monitoring, and continually improving soil health management systems and their outcomes; and (3) leveraging diverse partnerships and efforts across multiple organizations and geographical scales for further research and innovation in soil health assessment and management at local, regional, national, and global scales through standardized datasets and sharing information for agricultural lands. Having meaningful, science-based soil health assessments is also important for planning, implementing, and managing conservation projects, establishing baselines, and documenting soil property and process changes over time to quantify outcomes of such projects.

## Soil Health Indicators and Methods

Four main criteria have been developed by the soil health community of researchers, agricultural service providers, and practitioners to select indicators and methods for high-through-put soil test laboratories (Larson & Pierce, 1991; Mausbach & Seybold, 1998; Doran & Zeiss, 2000; Moebius et al., 2007; Norris et al., 2020):

1) Soil Health Indicator Effectiveness (short-term sensitivity to management, usefulness)
2) Production Readiness (ease of use, cost effectiveness for labs and producers)
3) Measurement Repeatability
4) Interpretability for agricultural management decisions (directionally understood, management influence known, regional potential ranges known, outcome thresholds).

These were developed using scientific literature and robust discussions in a series of workshops coordinated by the Farm Foundation and Noble Research Institute through the Soil Renaissance program between 2014 and 2016 (https://www.farmfoundation.org/projects/the-soil-renaissance-knowledge-to-sustain-earths-most-valuable-asset-1873-d1/). Understanding that soil health is a dynamic and evolving component of soil science, we recognize that both the indicators and methods recommended within these two volumes could change. Potential factors leading to changes may include identification of: (1) new or different critical soil processes, (2) more-responsive SH indicators, and/or (3) better methods of assessment. Furthermore, because of the dynamic nature of soil health assessment, we suggest information in these volumes be reviewed in three to 5 yr or a decade at most.

## Need for Standardization

Once a suite of soil health indicators has been selected, standard methods for collecting and handling samples in the field, processing them in the laboratory, analyzing them, and interpreting the data are needed for monitoring and making appropriate comparisons (Doran & Parkin, 1994). This is especially true for biological assays which can be more sensitive to how soil samples are collected and processed prior to analysis than to subtle differences in the analytical methods themselves. Currently, soil health measurement protocols vary widely and can therefore lead to inconsistent results and slow progress toward widely validated interpretation. This challenge is best addressed by standardization of a minimum dataset of methods used across organizations that collaborate nationally to make progress on interpretation and science-based management recommendations (USDA-NRCS, 2019b). Thus, ongoing efforts among public-sector and commercial laboratories are needed to ensure preanalytical soil processing (i.e., degree of aggregation, sieving, grinding, etc.) and analytical methods are standardized. As with all soil chemical measurements (e.g., pH, salinity, extractable N, phosphorus, and potassium), biological and physical indicators generally have large spatial and temporal variation. Care thus needs to be taken not only with sampling (i.e., compositing enough subsamples to make inferences about a sampled area) but also sampling methods (soil volume and depth), timing of collection (seasonal or annual), and the statistical methods used for interpretation.

Volume 2 is also intended to help reduce analytical variation in the measurement of soil health indicators. This is important because, as previously shown by the standardization of NRCS inherent soil property characterization methods, standardization makes large-scale data integration and comparisons feasible. Without rigorous standardization of soil health methods, variation among laboratories will hinder evaluation of changes over time and space and development of interpretations for various soil types and climate scenarios. This will in turn make regional and national compilations of soil health data very difficult to interpret.

Standardization of methods and protocols, along with appropriate proficiency testing, will facilitate collection of high-quality data with a high degree of interpretability, which is needed to facilitate development and use of regionally-appropriate interpretation functions (*i.e.*, scoring algorithms). Those algorithms are needed to transform raw laboratory data into unitless (0 to 1) values that shows how well a specific soil is performing a production or environmental function. Such ratings can then be used for on farm management decision making. Private and public soil testing laboratories that use broadly standardized methods will therefore have the advantage of being able to offer broadly validated soil health testing and interpretation using functions and recommendations developed from a large dataset achieved through multiorganization public-private partnership contributions.

## Interpretation of Soil Health Information

Several nationally appropriate tools, including the Revised Universal Soil Loss Equation (RUSLE), Soil Conditioning Index (SCI), Water Erosion Prediction Project (WEPP), Wind Erosion Prediction System (WEPS), AgroEcosystem Performance Assessment Tool (AEPAT), and Soil Management Assessment Framework (SMAF), have been developed to help interpret soil health related data (USDA-NRCS, 2019b). RUSLE2 estimates soil loss due to rill and inter-rill erosion caused by rainfall on cropland (Renard et al., 2011; USDA-ARS, 2015). The SCI combines information from the soil tillage intensity rating tool (STIR), a N-leaching index, and Version 2 of the Revised Universal Soil Loss Equation (RUSLE2) to provide information to producers regarding how their management decisions are affecting their soil resources and is widely used in NRCS conservation planning. AEPAT is a research-oriented index methodology that ranks agroecosystem performance among management practices for chosen functions and indicators (Liebig et al., 2004; Wienhold et al., 2006). Water Erosion Prediction Project (WEPP) is a process-based, distributed parameter, continuous simulation, erosion prediction model for use on personal computers (USDA-ARS, 2017); Wind Erosion Prediction System (WEPS) predicts many forms of soil erosion by wind including saltation-creep and suspension (USDA-ARS, 2018). Without question, wind-, water-, and anthropogenic-induced soil erosion continues to be a global problem (Karlen & Rice, 2015) and must be the first factor mitigated to truly improve soil health, as it is an advanced symptom of degradation including loss of soil organism habitat, stable aggregation, and other critical soil functions.

Soil health indicator measurements, when coupled with an available assessment framework, complement soil erosion tools as they can directly and more definitively detect less advanced symptoms of soil health degradation across diverse management systems. Laboratory data, without field-level information can be difficult to interpret or use for management decisions, and should only be used when supplemented with qualitative, in-field assessments of SH and an understanding of the past and current management system in use.

Data collected over time from the same field can be used to monitor soil health, but this may take a long time to be of value to producers or organizations, as it requires establishing a baseline and sampling over a number of years. Use of soil health assessment frameworks allow single field indicator measurements to be interpreted and used for decision making by leveraging a wealth of research conducted over the last 50 yr and continued targeted data collection. The first such framework (SMAF;) was developed collaboratively between ARS and NRCS (Andrews et al., 2004). Stott et al. (2010) and Wienhold et al. (2009) improved the SMAF by providing additional indicator scoring curves, thus improving its utility for both crop and pasture lands. SMAF uses broad soil taxonomic groups (suborders) as a foundation for assessment and allows curve modification based on inherent soil suborder characteristics. This is often essential as a contextual basis for indicator interpretation.

By design, SMAF assessments are soil- and site-specific, because they depend on soil, climate, and human values such as intended land use, management goals, and environmental sensitivity. A purported SMAF strength is that all of those factors can be manipulated by the user (primarily researchers). This will cause subtle changes in the scoring curves, causing some to argue that is not an advantage because it makes the process too complex for producers and their service providers. The approach taken by the SMAF was thereafter adapted for high throughput, public laboratory soil health testing in New York State by Idowu et al. (2008). The Comprehensive Assessment of Soil Health (CASH) was designed to evaluate soil functioning with respect to crop production and environmental impact and provide producers with a soil health status report similar to soil fertility reports commonly provided by soil testing labs. Most scores are effectively percentile ratings, comparing a measured value to the known population distribution in a soil textural group. CASH was based on the SMAF but used indicator methods with faster analytical procedures to accommodate a high-throughput lab setting. Furthermore, CASH was originally developed solely for New York, so scoring functions varied by texture, but were not adjusted for any other inherent soil characteristics associated with taxonomic classification or climate.

The framework approach for interpreting measured soil health data is further discussed in Volume 1 (Chapter 5). In summary, both SMAF and CASH provide efficient comparisons of similar soils under diverse management and estimates regarding the level of functioning of a particular field within the overall soil health continuum (van Es & Karlen, 2019). The key to robust interpretations is being able to compare soil samples from both agricultural and non-agricultural ecosystems, as well as for different soil and crop management practices, using consistent, standard, methods.

## Utilizing Soil Health Assessments to Inform Soil Management Decisions

It was stated in the Foreword to Doran et al. (1994) that "scientists and lay persons have long recognized that the quality of two great natural resources– air and water– can be degraded by human activity. Unfortunately, few people have considered that the quality of soil can also be affected by differing uses and management practices. Interest in *soil quality* has heightened during the past 3 yr as a small cadre of soil scientists became more concerned about the role of soils in sustainable production systems and the linkages between soil characteristics and plant-human health." This reflects just one early step in the exponential progress made during the past three decades that has led from soil quality being a research niche to broad awareness of the critical importance of healthy soils to agriculture and societies in general.

Soil health considerations are currently being incorporated across the activities of many agriculture-serving organizations nationally. For example, it has been incorporated into NRCS conservation planning and implementation programs. New soil health resource concerns, or constraints that can be documented by conservation planners, were published (USDA-NRCS, 2020a). These are also being embedded into key Conservation Practice Standards (USDA-NRCS, 2020b), a series of documents describing which constraints structural and agronomic conservation management practices can address, as well as the criteria for how to implement these practices to facilitate resolving identified constraints. Guidance is being updated to further reflect the current science on the importance of systems approaches as well, starting with crop lands, with plans to expand guidance available for other land uses. Standardization of measures is being undertaken across multiple organizations to build basic capacity to better inform management decisions and to quantify transition times and outcomes of soil health management systems implementation (USDA-NRCS, 2019b; Norris et al., 2020).

To fully address resource concerns and build fully functional soils, by improving organic matter quantity or quality, reversing soil organism habitat degradation, alleviating compaction, or improving soil aggregate stability, an agricultural annual cropping system that properly incorporates more than one soil health-targeted conservation practice is usually needed (Basche & DeLonge, 2017; Congreves et al., 2015; McDaniel et al., 2014; Turmel et al., 2015; USDA-NRCS, 2019a). Through the NRCS conservation planning process, conservation practices (USDA-NRCS, 2020b) for cover crops (340), crop rotation (328), and reduced- or no-tillage (345 and 329) are regularly employed by conservation planners to address various resource concerns since the methods of soil and crop management they represent are important contributors to sustainable agricultural production systems. However, if used in isolation, it is unlikely that a single practice will provide lasting SH benefits through improved SOM or the critical soil functions associated with SOM. For example, incorporating a cereal rye cover crop into a two-crop rotation that is dependent on frequent tillage may provide some weed suppression benefits or help reduce erosion, but it is unlikely that soil aggregate stability will be improved. Realizing improvement in soil function through management changes will typically require the strategic and simultaneous use of multiple practices. To better understand potential interactions among practices it may be helpful to think of their roles with regard to each of four broadly applicable NRCS soil health principles: minimizing soil disturbance, maximizing soil cover, maximizing biodiversity, and maximizing the presence of living roots. Each principle can be implemented using soil and crop management practices designed to address existing soil health concerns and maintain soil function. Depending on the cropping system, each soil health principle can be achieved through appropriate use of one or more conservation practices. Achieving all four principles by thoughtfully implementing and adaptively integrating multiple, complimentary conservation practices known to address identified constraints or concerns is the best way to ensure that soil health constraints are alleviated through synergistic effects as illustrated below.

### Minimizing Soil Disturbance

In some cropping systems, physical, chemical, or biological soil disturbance is an inevitable consequence of crop production (Schjønning et al., 2004). However, advances in agronomic research, farm equipment design, and technology have created the potential for most annual cropland acres to be managed with reduced- or often no-tillage practices. Inappropriate use of nutrients and pesticides can also cause soil ecosystem disturbance (Ellert et al., 1997; Frey et al., 1999; West & Post, 2002). Reducing disturbance helps slow carbon losses, minimizes physical destruction of aggregates, and maintains habitat for soil organisms (Larson et al., 1994). In addition to reduced- or no-tillage (345 and 329), Conservation Cover (327), IPM (595), Nutrient Management (590), and Prescribed Grazing (528) can also be implemented to minimize soil disturbance.

#### Maximizing Soil Cover

Crop residue and other organic materials such as mulch and compost, when they are left on the soil surface, provide a protective barrier between the soil and the destructive force of raindrops. They also moderate extremes in soil temperature and reduce evaporative losses from the soil. Soil cover can also be provided by leaves of growing plants. Keeping the soil covered throughout the year helps maintain soil aggregate integrity, protect habitat and provide food for soil organisms. Conservation practices that can be used to maximize cover include Conservation Cover (327), Cover Crop (340), Forage & Biomass Planting (512), Mulching (484), Prescribed Grazing (528) and Residue/Tillage Management (329/345).

#### Maximizing Biodiversity

It is well known that crop rotations are an important tool for managing plant pests (Altieri, 1991a, 1991b). What has been less appreciated until recently is that plants, primarily through their roots, affect the kinds and abundance of soil microorganisms, thus influencing soil biology and biological processes (Doran & Zeiss, 2000). Different plant species, and even cultivars, are typically associated with distinct soil microbial communities (Dick, 1997). In addition, since plant root architecture often differs among species, effects on soil function are also different (Brussaard et al., 2004). Above ground plant and animal diversity also encourages diversity in soil biology by increasing SOM levels, providing food and habitat for diverse soil communities, promoting greater aggregate stability, and helping alleviate compaction. Conservation practices that can be used to maximize biodiversity include Conservation Cover (327), Conservation Crop Rotation (328), Cover Crop (340), Forage & Biomass Planting (512), and Prescribed Grazing (528).

#### Maximizing the Presence of Living Roots

The area around plant roots is typically where the highest number and greatest diversity of soil microorganisms are found (Hornby & Bateman, 1997; Grayston et al., 1998; Ladygina & Hedlund, 2010; Singh et al., 2004). The rhizosphere is a very important ecological zone for SH improvement because living plant roots exude numerous carbon compounds as they grow and steadily slough dead cells from their surfaces. These contributions from roots add organic carbon to the ecosystem and help feed soil organisms. Plant roots are also involved in complex biochemical communication among soil microbes whereby beneficial organisms are "recruited" by plants while pathogenic organisms are often deterred. Plant roots also physically enmesh soil particles thus helping to create and preserve soil aggregates. Conservation practices that can be used to maximize the presence of living roots in the soil include Conservation Cover (327), Conservation Crop Rotation (328), Cover Crop (340), Forage & Biomass Planting (512), Mulching (484) and Prescribed Grazing (528). These selected practices are just some of the ways soil biological processes can be enhanced by SH management systems. Producers and those who work with them on establishing new management adaptations continue to innovate. The remaining chapters provide additional information and examples illustrating how continued implementation of known soil health promoting practices and new innovations can be assessed for their effects on critical soil functions by measuring appropriate SH indicators.

## Summary and Conclusion

Efforts to build agricultural resilience through high functioning soil resources are still in their infancy, as documented by national adoption rates for soil health associated practices, and especially soil health management systems across entire human-managed landscapes (Karlen & Rice, 2015; Wade et al., 2015). Fortunately, federal, state, NGO and private-sector organizations and individuals are working diligently to advance awareness of soil health and the management practices that improve it. Through increased research, on farm implementation, and policy changes progress is inevitable. The focus in "Approaches to Soil Health Analysis" is to build standardized, basic capacity to better inform management decisions and quantify outcomes of soil health management system implementation.

Soil health developments during the past three decades have been progressive, provocative, and are thus still under debate. As such, this two-volume contribution in no way is conceived as providing any final answers, but is envisioned as a step toward incorporating soil health into mainstream soil, water, and environmental science programs, and more importantly into every day agricultural management. Hopefully, they will also open new doors and stimulate additional study and education needed to encourage humankind to recognize the truth in Larson's often quoted statement that soil is "the thin layer covering the planet that stands between us and starvation" (Karlen et al., 2014).

## References

Acton, D. F., & Gregorich, L. J. (1995). The health of our soils: Toward sustainable agriculture in Canada. Pub. No. 1906/E Centre for Land and Biological Resources Research. Ottawa: Agriculture and Agri-Food Canada. https://doi.org/10.5962/bhl.title.58906

Alexander, M. (1971). Agriculture's responsibility in establishing soil quality criteria. In *Environmental improvement: Agriculture's challenge in the seventies* (pp. 66–71). Washington, DC: National Academy of Sciences.

Altieri, M. A. (1991a). How best can we use biodiversity in agroecosystems. *Outlook Agricultre*, 20, 15–23. https://doi.org/10.1177/003072709102000105

Altieri, M. A. (1991b). Increasing biodiversity to improve insect pest management in agro-ecosystems. In D. L. Hawksworth (Ed.), *The biodiversity of microorganisms and invertebrates: Its role in sustainable agriculture* (pp. 165–182). UK: CAB International.

Andrén, O., & Balandreau, J. (1999). Biodiversity and soil functioning: From black box to can of worms? *Applications in Soil Ecology*, 13, 105–108. https://doi.org/10.1016/S0929-1393(99)00025-6

Andrews, S. S., Karlen, D. L., & Cambardella, C. A. (2004). The soil management assessment framework: A quantitative soil quality evaluation method. *Soil Science Society of America Journal*, 68, 1945–1962. https://doi.org/10.2136/sssaj2004.1945

Basche, A., & DeLonge, M. (2017). The impact of continuous living cover on soil hydrologic properties: A meta-analysis. *Soil Science Society of America Journal*, 81(5), 1179–1190. https://doi.org/10.2136/sssaj2017.03.0077

Boehm, M., & Burton, S. (1997). Socioeconomics in soil-conserving agricultural systems: Implications for soil quality. In E. G. Gregorich & M. R. Carter (Eds.), *Soil quality for crop production and ecosystem health* (pp. 293–312). Amsterdam, Netherlands: Elsevier Science Publishers. https://doi.org/10.1016/S0166-2481(97)80040-2

Brussaard, L., Kuyper, T. W., Didden, W. A. M., de Goede, R. G. M., & Bloem, J. (2004). Biological soil quality from biomass to biodiversity – Importance and resilience to management stress and disturbance. In P. Schjønning, S. Elmholt, & B. T. Christensen (Eds.), *Managing soil quality: Challenges in modern agriculture* (pp. 139–161). Wallingford, U.K: CABI Publishing.

BASF. 2020. Poncho Votivo 2.0 Seed Treatment System. https://agriculture.basf.us/crop-protection/products/poncho-votivo-2-0.html (verified 13 July 2020).

Carter, M. C., Gregorich, E. G., Anderson, D. W., Doran, J. W., Janzen, H. H., & Pierce, F. J. (1997). Concepts of soil quality and their significance. In E. G. Gregorich & M. R. Carter (Eds.), *Soil quality for crop production and ecosystem health* (pp. 1–19). Amsterdam, Netherlands: Elsevier Science Publishers. https://doi.org/10.1016/S0166-2481(97)80028-1

Chendev, Y. G., Sauer, T. J., Ramirez, G. H., & Burras, C. L. (2015). History of East European Chernozem soil degradation: Protection and restoration by tree windbreaks in the Russian Steppe. *Sustainability*, 7(1), 705–724. https://doi.org/10.3390/su7010705

Congreves, K. A., Hayes, A., Verhallen, E. A., & Van Eerd, L. L. (2015). Long-term impact of tillage and crop rotation on soil health at four temperate agroecosystems. *Soil Tillage Research*, 152, 17–28. https://doi.org/10.1016/j.still.2015.03.012

DeLong, C., Cruse, R., & Wiener, J. (2015). The soil degradation paradox: Compromising our resources when we need them the most. *Sustainability*, 7(1), 866–879. https://doi.org/10.3390/su7010866

Dick, R. P. (1997). Soil enzyme activities as integrative indicators of soil health. In C. Pankhurst, B. M. Doube, & V. V. S. R. Gupta (Eds.), *Biological indicators of soil health* (pp. 121–156). New York: CAB International.

Doran, J. W., Coleman, D. C., Bezdicek, D. F., & Stewart, B. A. (Eds.) (1994). *Defining soil quality for a sustainable environment*. SSSA Special Publication No. 35. Madison, WI: SSSA.

Doran, J. W., & Parkin, T. B. (1994). Defining and assessing soil quality. In J. W. Doran, D. C. Coleman, D. F. Bezdicek, & B. A. Stewart (Eds.), *Defining soil quality for a sustainable environment* (pp. 3–21). SSSA Special Publication No. 35. Madison, WI: SSSA. https://doi.org/10.2136/sssaspecpub35

Doran, J. W., & A. J. Jones (Eds.) (1996). *Methods for assessing soil quality*. SSSA special Publication No. 49. Madison, WI: SSSA.

Doran, J. W., & M. R. Zeiss. (2000). Soil health and sustainability: Managing the biotic component of soil quality. *Applied Soil Ecology*, 15, 3–11. https://doi.org/10.1016/S0929-1393(00)00067-6

Doran, J. W. (2002). Soil health and global sustainability: Translating science into practice. *Agriculture Ecosystems Environment*, 88(2), 119–127. https://doi.org/10.1016/S0167-8809(01)00246-8

Ellert, B. H., Clapperton, M. J., & Anderson, D. W. (1997). An ecosystem perspective of soil quality. In E. G. Gregorich & M. R. Carter (Eds.), *Soil quality for crop production and ecosystem health* (pp. 115–141). Elsevier Amsterdam, Netherlands: Science Publishers. https://doi.org/10.1016/S0166-2481(97)80032-3

Frey, S. D., Elliott, E. T., & Paustian, K. (1999). Bacterial and fungal abundance and biomass in conventional and no-tillage agroecosystems along two climatic gradients. *Soil Biology & Biochemistry*, 31(4), 573–585. https://doi.org/10.1016/S0038-0717(98)00161-8

Gelaw, A. M., Singh, B. R., & Lal, R. (2015). Soil quality indices for evaluating smallholder agricultural land uses in northern Ethiopia. *Sustainability*, 7(3), 2322–2337. https://doi.org/10.3390/su7032322

General Mills. (2020). Regenerative Agriculture. https://www.generalmills.com/Responsibility/Sustainability/Regenerative-agriculture (verified 13 July 2020).

Glæsner, N., Helming, K., & de Vries, W. (2014). Do current European policies prevent soil threats and support soil functions. *Sustainability*, 6(12), 9538–9563. https://doi.org/10.3390/su6129538

Grayston, S. J., Wang, S., Campbell, C. D., & Edwards, A. C. (1998). Selective influence of plant species on microbial diversity in the rhizosphere. *Soil Biology & Biochemistry*, 30(3), 369–378. https://doi.org/10.1016/S0038-0717(97)00124-7

Gregorich, E. G., & Carter, M. R. (Eds.) (1997). *Soil quality for crop production and ecosystem health*. Amsterdam, Netherlands: Elsevier Science Publishers.

Havlicek, E., & Mitchell, E. A. (2014). Soils supporting biodiversity. In J. A. Krumins & J. Dighton (Eds.), *Interactions in soil: Promoting plant growth* (pp. 27–58). Springer, Dordrecht. https://doi.org/10.1007/978-94-017-8890-8_2

Herrick, J. E., Brown, J. R., Bestelmeyer, B. T., Andrews, S. S., Baldi, G., Davies, J., Duniway, M., Havstad, K. M., Karl, J., Karlen, D. L., Peters, D. P. C., Quinton, J. N., Riginos, C., Shaver, P. L., Steinaker, D., & Twomlow, S. (2012). Revolutionary land use change in the 21st century: Is (rangeland) science relevant? *Rangeland Ecology & Management*, 65, 590–598. https://doi.org/10.2111/REM-D-11-00186.1

Hillel, D. (1991). *Out of the earth: Civilization and the life of the soil*. Oakland, CA: University of California Press.

Hornby, D., & Bateman, G. L. (1997). Potential use of plant root pathogens as bioindicators of soil health. In C. Pankhurst, B. M. Doube, & V. V. S. R. Gupta (Eds.), *Biological indicators of soil health* (pp. 179–200). New York: CAB International.

Idowu, O. J., van Es, H. M., Abawi, G. S., Wolfe, D. W., Ball, J. I., Gugino, B. K., Moebius, B. N., Schindelbeck, R. R., & Bilgili, A.V. (2008). Farmer-oriented assessment of soil quality using field, laboratory, and VNIR spectroscopy methods. *Plant & Soil*, 307, 243–253. https://doi.org/10.1007/s11104-007-9521-0

Karlen, D. L., Mausbach, M. J., Doran, J. W., Cline, R. G., Harris, R. F., & Schuman, G. E. (1997). Soil quality: A concept, definition, and framework for evaluation. *Soil Science Society of America Journal*, 61, 4–10. https://doi.org/10.2136/sssaj1997.03615995006100010001x

Karlen, D. L., Peterson, G. A., & Westfall, D. G. (2014). Soil and water conservation: Our history and future challenges. *Soil Science Society of America Journal*, 78, 1493–1499. https://doi.org/10.2136/sssaj2014.03.0110

Karlen, D. L., & Rice, C.W. (2015). Soil degradation: Will humankind ever learn? *Sustainability*, 7, 12490–12501. https://doi.org/10.3390/su70912490

Lal, R. (2015). Restoring soil quality to mitigate soil degradation. *Sustainability*, 7, 5875–5895. https://doi.org/10.3390/su7055875

Larson, W. E., & Pierce, F. J. (1991). Conservation and enhancement of soil quality. Evaluation for sustainable land management in the developing world. In IBSRAM Conference Proc. 12th. Int. Board for Soil Research and Management, Jatujak Thailand, Bangkok, Thailand. p. 175–203.

Larson, W. E., Eynard, A., Hadas, A., & Lipiec, J. (1994). Control and avoidance of soil compaction in practice. In B. D. Soanee & C. van Ouwerkerk (Eds.), *Soil compaction in crop production* (pp. 597–625). Developments in Agricultural Engineering 11. Amsterdam, Netherlands: Elsevier. https://doi.org/10.1016/B978-0-444-88286-8.50033-7

Ladygina, N., & Hedlund, K. (2010). Plant species influence microbial diversity and carbon allocation in the rhizosphere. *Soil Biology & Biochemistry*, 42(2), 162–168. https://doi.org/10.1016/j.soilbio.2009.10.009

Liebig, M. A., Miller, M. E., Varvel, G. E., Doran, J. W., & Hanson, J. D. (2004). AEPAT: Software for assessing agronomic and environmental performance of management practices in long- term agroecosystem experiments. *Agronomy Journal*, 96, 109–115. https://doi.org/10.2134/agronj2004.0109

Mausbach, M. J., & Seybold, C. A. (1998). Assessment of soil quality. In R. Lal (Ed.), *Soil quality and agricultural sustainability* (pp. 33–43). Chelsea, MI: Ann Arbor Press.

McDaniel, M. D., Tiemann, L. K., & Grandy, A. S. (2014). Does agricultural crop diversity enhance soil microbial biomass and organic matter dynamics? A meta-analysis. *Ecological Applications*, 24(3), 560–570. https://doi.org/10.1890/13-0616.1

Moebius, B. N., van Es, H. M., Schindelbeck, R. R., Idowu, O. J., Thies, J. E., & Clune, D. J. (2007). Evaluation of laboratory-measured soil properties as indicators of soil physical quality. *Soil Science*, 172, 895–912. https://doi.org/10.1097/ss.0b013e318154b520

Montgomery, D. R. (2007). *Dirt: The erosion of civilizations*. Berkley, CA: University of California Press.

National Research Council. (1993). *Soil and water quality: An agenda for agriculture*. Washington, DC: National Academic Press.

Norris, C. E., Bean, G. M., Cappellazzi, S. B., Cope, M., Greub, K. L. H., Liptzin, D., Rieke, E. L., Tracy, P. W., Morgan, C. L. S., & Honeycutt, C. W. (2020). Introducing the North American project to evaluate soil health measurements. *Agronomy Journal*. https://doi.org/10.1002/agj2.20234

Nutrien Ltd. (2018). Nutrien Announces Agreement to Purchase Waypoint Analytical, a Leading U.S. Agricultural Lab and Soil Science Company. https://www.prnewswire.com/news-releases/nutrien-announces-agreement-to-purchase-waypoint-analytical-a-leading-us-agricultural-lab-and-soil-science-company-300677498.html (verified 13 July 2020).

Pankhurst, C. E., Doube, B. M., & Gupta, V. V. S. R. (1997). Biological indicators of soil health: Synthesis. In C. E. Pankhurst, B. M. Doube, & V. V. S. R. Gupta (Eds.), *Biological indicators of soil health* (pp. 419–435). New York: CAB International.

Paustian, K., Lehmann, J., Ogle, S., Reay, D., Robertson, G. P., & Smith, P. (2016). Climate-smart soils. *Nature*, 532(7597), 49–57. https://doi.org/10.1038/nature17174

Renard, K. G., Yoder, D. C., Lightle, D. T., & Dabney, S. M. (2011). Universal soil loss equation and revised universal soil loss equation. In R. P. C. Morgan & M. A. Nearing (Eds.), *Handbook of erosion modelling* (pp. 137–167). Oxford, UK: Blackwell Publishing.

Schjønning, P., Elmholt, S., & Christensen, B. T. (2004). Soil quality management–Concepts and terms. In P. Schjønning, S. Elmholt, & B. T. Christensen (Eds.), *Managing soil quality: Challenges in modern agriculture* (pp. 1–15). Wallingford, UK: CABI Publishing.

Singh, B. K., Millard, P., Whiteley, A. S., & Murrell, J. C. (2004). Unravelling rhizosphere–microbial interactions: Opportunities and limitations. *Trends in Microbiology*, 12(8), 386–393. https://doi.org/10.1016/j.tim.2004.06.008

Stott, D. E., Andrews, S. S., Liebig, M. A., Wienhold, B. J., & Karlen, D. L. (2010). Evaluation of β- glucosidase activity as a soil quality indicator for the soil management assessment framework (SMAF). *Soil Science Society of America Journal*, 74, 107–119. https://doi.org/10.2136/sssaj2009.0029

Sustainability Reports. (2019). 86% of S&P 500 Index Companies Publish Sustainability / Responsibility Reports in 2018. https://www.sustainability-reports.com/86-of-sp-500-index-companies-publish-sustainability-responsibility-reports-in-2018/

Tilman, D. G. (1998). The greening of the green revolution. *Nature*, 396, 211–212. doi:10.1038/24254

Turmel, M. S., Speratti, A., Baudron, F., Verhulst, N., & Govaerts, B. (2015). Crop residue management and soil health: A systems analysis. *Agricultural Systems*, 134, 6–16. https://doi.org/10.1016/j.agsy.2014.05.009

USDA-ARS. (2015). Revised Universal Soil Loss Equation, Version 2 (RUSLE2). USDA National Soil Erosion Research Laboratory. https://data.nal.usda.gov/dataset/revised-universal-soil-loss-equation-version-2-rusle2 (verified 12 June 2020).

USDA-ARS. (2017). Water Erosion Prediction Project (WEPP). USDA Agricultural Research Service. Washington, DC. https://data.nal.usda.gov/dataset/water-erosion-prediction-project-wepp (verified 12 June 2020).

USDA-ARS. (2018). Wind Erosion Prediction System (WEPS). USDA ARS Engineering & Wind Erosion Research Unit. Washington, DC. https://data.nal.usda.gov/dataset/wind-erosion-prediction-system-weps (verified 12 June 2020).

USDA-NRCS. (2019a). The Basics of Addressing Resource Concerns with Conservation Practices within Integrated Soil Health Management Systems on Cropland, by D. Chessman, B.N. Moebius-Clune, B.R. Smith, and B. Fisher. Soil Health Technical Note No. 450-04. Available on NRCS Electronic Directive System. Washington, DC. https://directives.sc.egov.usda.gov (verified 12 June 2020).

USDA-NRCS. (2019b). Recommended Soil Health Indicators and Associated Laboratory Procedures, by D.E. Stott. Soil Health Technical Note No. 430-03.

Available on NRCS Electronic Directive System. Washington, DC. https://directives.sc.egov.usda.gov (verified 12 June 2020).

USDA-NRCS. (2020a). Electronic Directive, National Bulletin. NB 450-20-1 Field Office Technical Guide Resource Concerns and Planning Criteria List and Update (Title 450). Washington, DC. https://directives.sc.egov.usda.gov/ (verified 12 June 2020).

USDA-NRCS. (2020b). Electronic Directive, National Handbook of Conservation Practices (Title 450), Part 620, Subpart C, National Practice Standards. Washington, DC https://directives.sc.egov.usda.gov/(verified 6-12-2020).

USDA-NRCS. (2020c). NRCS Conservation Activity Soil Testing (Code 216). Washington, DC. https://directives.sc.egov.usda.gov/ (verified 12 June 2020).

van Es, H. M., & Karlen, D. L. (2019). Reanalysis confirms soil health indicator sensitivity and correlation with long-term crop yields. *Soil Science Society of America Journal*, 83, 721–732. https://doi.org/10.2136/sssaj2018.09.0338

Vitro, I., Imaz, M. J., Fernández-Ugalde, O., Gartzia-Bengoetxea, N., Enrique, A., & Bescansa, P. (2015). Soil degradation and soil quality in Western Europe: Current situation and future perspectives. *Sustainability*, 7(1), 313–365. https://doi.org/10.3390/su7010313

Wade, T., Claassen, R., & Wallander, S. (2015). Conservation-practice adoption rates vary widely by crop and region. Economic Information Bulletin (EIB) No.147, U.S. Department of Agriculture (USDA), Economic Research Service (ERS). Washington, DC.

Warkentin, B. P., & Fletcher, H. F. (1977). Soil quality for intensive agriculture. In Proceedings of the International Seminar on Soil Environment and Fertilizer Management in Intensive Agriculture. Society of Science of Soil and Manure, Japan. p. 594–598.

West, T. O., & Post, W.M. (2002). Soil organic carbon sequestration rates by tillage and crop rotation. *Soil Science Society of America Journal*, 66(6), 1930–1946. https://doi.org/10.2136/sssaj2002.1930

Wienhold, B.J., Pikul, J. L., Liebig, M. A., Mikha, M. M., Varvel, G. E., Doran, J. W., & Andrews, S.S. (2006). Cropping system effects on soil quality in the Great Plains: Synthesis from a regional project. *Renewable Agricultural Food Systems*, 21, 49–59. https://doi.org/10.1079/RAF2005125

Wienhold, B. J., Karlen, D. L., Andrews, S. S., & Stott, D.E. (2009). Protocol for soil management assessment framework (SMAF) soil indicator scoring curve development. *Renewable Agricultural Food Systems*, 24, 260–266. https://doi.org/10.1017/S1742170509990093

# 2

## Evolution of the Soil Health Movement

*Douglas L. Karlen, Mriganka De, Marshall D. McDaniel, and Diane E. Stott*

Soil Health, during the second decade of the 21st Century, has become a familiar term to both rural and urban audiences. Some may think the concept is new, but as outlined herein, the projects, workshops, books, and all other activities addressing this topic are built on a solid foundation reflecting numerous research, education, and technology contributions such as soil conservation, soil condition, soil tilth, soil carbon management, soil quality, soil security, or simply prevention of soil degradation. We have broken the evolution of soil health activities into four stages: (i) pre-20th Century contributions, (ii) soil tilth and conservation activities between ~1900 and 1970, (iii) introduction and initial soil quality activities, and (iv) acceptance, promotion, and adoption of soil health *per se*. Recognizing some contributions have been missed, we hope the presentation will provide a reasonable foundation for many different readers.

## Introduction

As the second decade of the 21st Century ends, the term "soil health" has become an accepted phrase, embedded globally in technical and non-technical writings. Federal and state government, non-government organizations, foundations, institutes, college and university curricula, public-private-partnerships, and numerous other entities have embraced the concept and thus embedded the term into the vernacular of many groups (Figure 2.1). For those who have spent recent decades striving to encourage adoption of soil health principles and the management practices required to implement them, global recognition and acceptance of soil health is gratifying, but we fully acknowledge that our small and humble

*Soil Health Series: Volume 1 Approaches to Soil Health Analysis*, First Edition.
Edited by Douglas L. Karlen, Diane E. Stott, and Maysoon M. Mikha.

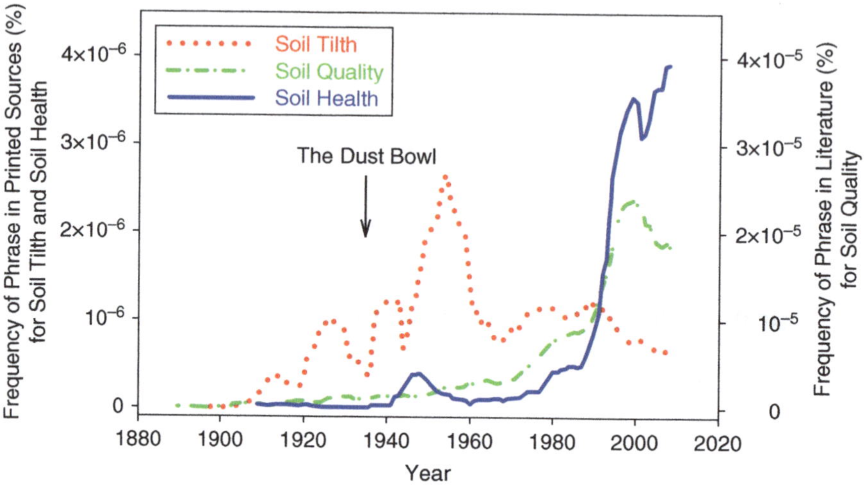


**Figure 2.1** An exponential increase in the use of soil quality and soil tilth in published literature. (Developed by M.D. McDaniel using https://books.google.com/ngrams [Michel et al., 2011]).

contributions were built on foundations laid by many before us. Some may regard the concept as new and unique, but soil health *per se* evolved from several soil management focus areas including soil condition, soil tilth, soil management, soil conservation, soil care, soil quality, soil productivity, soil resilience, soil security, and soil degradation.

Advocates for the care and wise use of soil have been warning humankind since before the common era (BCE) that soil (a.k.a. Land) is the foundation for everything we do or share (e.g., food security; water infiltration, retention and release; environmental buffering; biodiversity). Many pioneers, including H. H. Bennett who in response to the American Dust Bowl and many other improper soil management decisions successfully established the U.S. Soil Conservation Service (SCS), dedicated their lives to protecting and improving soils (Bennett, 1950). But, without question, something unique happened during the 1970s and 1980s (Figure 2.1) that spurred interest and resulted in an exponential increase in the words soil quality and soil health in titles, keywords, and abstracts from which literature search databases are built.

We cannot identify any single event (e.g., The Dust Bowl) or explain why the importance of soil resources was finally recognized and publically discussed. Perhaps, it was the establishment of "Earth Day" in 1970 which significantly increased awareness of unintended environmental impacts associated with post-World War II industrial developments (e.g., non-point pollution; surface- and ground-water contamination and depletion; wind-, tillage- and water-induced soil erosion; acid rain; increased emission of greenhouse gases).

None-the-less, soil resources, some 40 years after the U.S. Dust Bowl, were once again being recognized as fragile and in need of appropriate care and management to sustain them. During that era, our now deceased mentor, colleague, and friend, W. E. (Bill) Larson, often described soil as "the thin layer covering the planet that stands between us and starvation" (Karlen et al., 2014a). This quote parallels writings by two other soil science pillars whom we suggest indirectly helped lay a foundation for soil health. The first is W.C. Lowdermilk (1953), who summarized his personal experiences in 1938 and 1939 in an often-reproduced publication entitled "Conquest of the Land through 7,000 Years." His writings began by recognizing that human civilizations literally wrote their records on the land. He used those experiences to help raise public awareness of soil erosion problems occurring within the United States and globally. The second is Hillel (1991) who in his book "Out of the Earth: Civilization and the Life of the Soil" included a treatise in which Plato has Critias deliver a proclamation comparing degraded land and soil resources to an abandoned, emaciated person.

Science-based principles influencing or even controlling overall soil health and the critical functions healthy soils provide for humankind can be traced to Aristotle or van Helmont, who provided some of the first insight and understanding of how plants obtain their nutrients from soils. Carter et al. (1997) quoted Columella, a prominent writer about agriculture within the Roman Empire, to illustrate his early (~40 to 60 BCE) guidance on how soil resources differ, the impact of terrain or landscape position, and how to manage them using virtues of good soil husbandry [management], such as optimizing soil moisture status, incorporating plant residues and/or manure to restore ("grow fat") the soil. Bennett (1950) in his report on American Land emphasized that "*we can't keep our present standard of living if we lose much more* [soil]." He continued stating that "*more waste of good land would amount to a national crime on the part of those who are responsible – meaning ourselves.*"

Another soil science pioneer, Hans Jenny (1980) stressed interrelationships between soil type and soil properties, while in the 1990s Warkentin (1995) emphasized how tillage energy as well as irrigation, drainage, and fertilizer inputs are all factors affecting the quality of soil for crop production. Carter et al. (1997) have also provided many details regarding the history and evolution of the soil quality concept, which we consider to closely parallel those of soil health and thus use the terms interchangeably throughout these two volumes. Carter et al. (1997) does, however, provide excellent insight regarding subtle differences between the two terms. The soil quality concept emerged as efforts were made to place a value on soil resources for providing specific functions, serving a specific purpose, or supporting a specific use. But, in contrast to water and air quality for which their functions can be directly related to human or animal consumption, soil functions are generally more diverse and usually cannot be directly linked to human health.

One of the few situations that do closely link soil and human health is the function of filtering and buffering, especially if the heavy metals, radionuclotides, and/or toxic organic compounds can enter surface and groundwater resources or the food chain. This issue, specifically with regard to lead (Pb) is addressed in Chapter 7 of this volume.

More recently, Jared Diamond (2005) and David Montgomery (2007) inspired public awareness of the fragility of soil resources in their books entitled "Collapse: How Societies Choose to Fail or Succeed" and "Dirt: The Erosion of Civilizations", respectively. Building on the foundations laid by those writers and numeerous agricultural scientists and engineers, our objective for this chapter is to briefly summarize and acknowledge the insight, passion, and selected contributions that have contributed to soil health evolution during the two decades since publication of the soil quality books edited by Doran et al (1994) and Doran and Jones (1996).

## Pre-20$^{th}$ Century Soil Awareness

Soil health is built upon a solid foundation reflecting numerous agronomic and soil science publications and advancements in knowledge. We would be remise not to mention classic scientists, like Darwin and Dukochaev, that spearheaded not only the modern scientific revolution but also our understanding of soil development and biology (e.g., Brevik and Hartemik, 2010; Johnson and Schaetzl, 2015; Ghilarov, 1983). Likewise, it's imparitive to acknowledge the indigenous ecological knowledge on soil health acquired by non-colonial cultures over thousands of years (Pawluk, 1992; Raji, 2006). Those roots of soil health are critical, but beyond the focus of this chapter. Therefore, we start with the late 20$^{th}$ Century and in the United States, where the concept of tilth (Warkentin 2008; Karlen 1990) is considered a key driver for what is now recognized as soil health. One example, provided in a review of past, present, and future soil tilth issues (Karlen et al. 1990) is a 1523 book by Fitzherbert, entitled "Boke of Husbandry." Therein, Fitzherbert wrote that to grow peas (*Pisum sativum* L.) or beans (*Phaseolus vulgaris* L.) the soil was not ready to be planted "if it synge or crye, or make any noise under thy fete" whereas "if it make no noyse and wyll beare they horses, thane sowe in the name of God" (Keen, 1931). A similar quote from Fream (1890) described "good" soil as being "open, free-working, mellow or in good heart" and "non-productive" soil as "hungry, stubborn, still, cold, or unkind." For some, these descriptions of "good" and "non-productive" soils cause them to recall the Biblical connection between soil and life described by Hillel (1991) or even the New Testament parable wherein Jesus describes people and their actions as being similar to one for the four soil categories.

## From 1900 to 1970

Examples of influential publications include one by Yoder (1937), who concluded poor soil structure was a major factor limiting various soil functions because of its influence on processes including granulation (aggregation); wetting, drying, freezing and thawing cycles; organic matter accumulation and decomposition rates; biological activities, plant root development, as well as tillage and crop rotation response. This remains relevant because the work led to development of the "Yoder" water-stable aggregate method that is currently being used for many soil health assessment projects. Another is Waksman and Starkey (1924) who measured $CO_2$ emissions as an indicator of the "decomposing power of soils" – using procedures similar to those associated with the Haney test (Haney et al., 2006).

Wilson and Browning (1945) emphasized soil aggregation and documented significant differences for a corn (*Zea mays* L.) – oat (*Avena sativa* L.) - meadow rotation versus continuous corn. They also reported that after only four years of continuous corn aggregation levels created by either alfalfa (*Medicago sativa* L.) or bluegrass (*Poa pratensis* L.) was decreased to levels below that found for the corn-oat-meadow rotation. Conversely, implementing the corn-oat-meadow rotation after 11 years of continuous corn increased aggregation. Another pioneer was Selma Waksman, who is more famous for the discovery of streptomycin, conducted early work on understanding the nature of soil organic matter (SOM), including the decomposition of plant components by soil microorganisms and preservation of nitrogen (N) in the SOM (Waksman and Hutchings, 1935). Waksman and colleagues also focused on the formation of soil aggregates during the microbial decomposition process (Martin and Waksman, 1939, 1941). In another review, Whiteside and Smith (1941) documented the importance of measuring SOM and total N, two factors important for soil health. They concluded that since the origin of agriculture *per se* soil productivity had gradually been depleted due to crop production. Similarly, Van Bavel and Schaller (1950) showed that both cropping systems and soil erosion influenced SOM content. They reported that 11 years of continuous alfalfa increased SOM, but 11 years of bluegrass did not. They also found that although a corn-oat-meadow rotation did not maintain SOM levels, although the decrease was "small and probably not significant."

Post-World War II soil management studies gradually began to focus more on soil physical and chemical manipulation than biology, primarily because equipment manufacturing and fertilizer development technologies advanced rapidly, while biological theories (e.g., discovery of deoxyribonucleic acid [DNA]), methods of analysis, and modern instrumentation took additional decades to evolve. The importance of soil biology was not overlooked, as confirmed by Lyon et al (1950) who concluded plowing and cultivation should be used to loosen the soil with a minimum of soil aggregate destruction. Similarly, Browning and Norton

(1947) found that crop yield associated with moldboard plowing was generally better than yields associated with other forms of tillage. They attributed this response to the moldboard plow design, which was intended to accentuate granulation by lifting, twisting, and shearing the soil, while at the same time, inverting organic residues on or at the soil surface for subsequent decomposition. During this same era, advances in chemical weed control were also made, but the use of tillage was still favored despite writings such as "Plowman's Folly" by Faulkner (1943).

Tillage and cropping system studies such as those by Van Doren and Klingebiel (1952) and Klingebiel and O'Neal (1952) provided data that would now be regarded as critical soil health information. They reported that under virgin conditions, surface soil structure in most Corn Belt loam and silt loam soils was granular or crumb-like. The soils were also highly aggregated and generally had low bulk density. However, as tillage intensity increased and crop rotation diversity decreased, the granular soil structure deteriorated to a fine, fragmented or massive condition. Annual moldboard plowing, coupled with multiple near-surface tillage operations to prepare appropriate seedbeds and enhance oxidation of plant roots and other residues, was recommended, it ultimately resulted in decreased SOM levels, increased raindrop compaction, decreased infiltration, increased runoff, and greater soil erosion. Declining SOM also decreased plant available N, but post-World War II advances in chemical N fixation and availability of new fertilizer materials resulted in a gradual substitution of capital for labor and SOM. This transition was described by Melsted (1954) as "replacing the art of farming with the science of farming" (Karlen et al., 1990).

Without question, the "Father of Soil Conservation" Hugh Hammond Bennett focused public attention on the critical need for and merits of soil conservation. In his overview of our American land, he clearly articulated the perils of soil erosion and how it has resulted in humankind "wasting the gifts of nature." Continuing, Bennett clearly described the importance of productive land; emphasized that soil conservation is simply proper use and care of the land; outlined problems that soil conservation could help prevent; presented stratigies for conserving soil resources; and introduced Soil Conservation Districts (SCD) which began to be organized near his home in Anson County, NC, in 1937. It was through SCDs that farmers and land owners were introduced to detailed farm plans and new soil management practices (e.g. building of terraces, contour cultivation, soil testing, drainage, strip cropping, cover crops, increased perennialization, restoring woodlands and pasture, stubble mulching, and planting windbreaks). Differences in on-farm soil resources, fundamentally linked to inherent soil properties and processes, were articulated through land capability maps and the identification of Class I to Class VIII soils based on their potential and sustainability requirements. Farmers and soil conservationists were encouraged to work together to determine the best possible use for every part and parcel of each individual farm. Is this not what every soil health initiative is advocating?

Furthermore, since humankind had already ignored warnings from Plato, Aristotle, Columella, Fitzherbert, Fream, as well as an untold number of indigenous leaders regarding the care and management of our soil resources, ravages of the American Dust Bowl could likely have been predicted. However, a steadily increasing population, growing demand for food, feed, and fiber, advances in technology regarding how to till vast areas of the Great Plains, and well-intentioned, but inappropriate federal land use policies, coupled with farm economics due to the Great Depression, changes in regional weather patterns and other cultural factors contributed to the disaster (History.com Editors, 2020). Fortunately, due to science-based leadership by Yoder, Lowdermilk, Bennett and others, the U.S. Congress passed Public Law 74-46, which recognized "the wastage of soil and moisture resources on farm, grazing, and forest lands is a menace to the national welfare." The act created the USDA Soil Conservation Service (SCS), now known as the Natural Resources Conservation Service (NRCS), and helped promote many of the practices outlined by Bennett (1950), including sound land use, adherence to carrying capacity, and development of farm conservation plans.

With arrival of another prolonged drought in the 1950s, Congress passed the Great Plains Conservation Program which focused financial assistance for conservation in the Plains states. SCS provided financial and technical assistance to meet multiple objectives of conservation and economic stability. This included providing technical assistance for the Soil Bank Program (SBP), precursor to current Conservation Reserve Programs (CRP). The SBP paid to retire degraded cropland and provided financial incentives to farmers for planting protective cover crops (NRCS, 2018).

Conservation was given another boost under the Kennedy and Johnson administrations in the 1960s when SCS' role was expanded to address both rural and urban land use. Through an emphasis on rural development, SCS began to work with landowners in areas larger than small watersheds or conservation districts. The late 1960s can also be characterized as a time of broad popular concern regarding the environment, health, and welfare of people throughout the United States and around the world (e.g., Norman Borlaug and the Green Revolution). The first Earth Day in 1970 increased environmental awareness, and ultimately a national framework of environmental policies was created that changed the way the SCS put conservation on the ground. Soon, federal agencies were required to evaluate and report on the environmental impacts of their activities.

Water quality and non-point source pollution became important areas of concern. Protection of wetlands emerged as critical issue with SCS participation in the Water Bank program and provided incentives to landowners to protect wetland habitat (NRCS, 2018). During the 1970s, SCS also gained greater authority to monitor and assess the nation's natural resource base through the National Resources Inventory (NRI) – a focal point for future soil quality assessment studies. Finally, in response to these expanded authorities, the Congress changed SCS's name to the Natural Resources Conservation Service (NRCS) in 1994 (NRCS, 2018).

For brevity, this chapter references only a few significant soil research studies, laws and policies. For example, one federal program that did lead efforts to include soil health in the research portfolio for was the Sustainable Agricultue Research and Education (SARE)] program, but until recent private sector efforts (Chapter 1) steadily increasing farm size and purchased inputs minimized the impact of that program across the broader agricultural community. This chapter has also likely failed to appropriately recognize many of the pillars in soil management research and extension (Karlen et al., 2014a) that contributed substantially to the soil conservation foundation upon which soil health has evolved. Some of those outstanding contributions include studies by Balfour, Bidwell, Hole, Hyams, Jenny, Leopold, Rodale, Whitney, Yaalon, and many others. Without question, all contributed significant knowledge and understanding the soil functions providing the science-based foundation for current soil health endeavors.

## From 1970 to 2000 – Soil Quality Emerges

### Concept Development

As noted previously, soil physical and chemical aspects of soil management dominated post-World War II activities for most soil and crop scientists. We suggest this reflected our limited understanding of soil biological properties and processes as well as the lack of instrumentation and analytical tools that are now available. None-the-less several well recognized soil microbiologists such as Allison (1968, 1973), James P. Martin (1939, 1940, 1971), Eldor Paul (Mathur and Paul, 1967; Paul and Voroney, 1980; Paul, 2014), and Martin Alexander (Alexander, 1961, 1980; Acea et al., 1988) contributed insights that expanded the foundation upon which soil health has evolved. There was also an increasing awareness that decreased use of crop rotations, increased size and weight of farm tractors and implements, as well as increased use of conservation tillage practices, were having measurable soil tilth effects in the northern Corn Belt (Voorhees, 1979). Soil compaction, caused by those factors, was recognized as being important for several reasons, including its effect on annual freezing and thawing processes (Voorhees et al., 1978; Voorhees, 1983; Voorhees and Lindstrom, 1984). Coupled with increasing concern regarding soil degradation, Pierce et al. (1983, 1984) conducted several erosion – productivity studies and Dormaar et al. (1988)

intentionally eroded a Dark Brown Chernozemic soil (Mollisol) to demonstrate that applying commercial fertilizer or manure could restore soil productivity. However, during a subsequent drought, only the sites that received manure maintained yields and furthermore, after five years those treatments showed increased SOM content and improved water-stable aggregation.

Warkentin and Fletcher (1977) introduced the concept of soil quality, which in many ways became the foundation for current soil health activities. The transition to soil quality emphasized the multiple ecosystem services (i.e., food and fiber production, recreation, and recycling or assimilation of wastes or other by-products) that soils must provide (Carter et al., 1997). A focus on soil quality required recognition that: (1) soil resources are constantly being evaluated for many different uses; (2) multiple stakeholder groups are concerned about soil resources; (3) society's priorities and demands on soil resources are changing; and (4) soil resource and land use decisions are made in a human or institutional context (Warkentin and Fletcher, 1977). They also stated that because of inherent differences among soils, there is no single measurement that will always be useful for evaluating soil quality (Karlen et al., 2003a).

Another 1980s soil and crop management challenge influencing SOM, erosion, and crop productivity was the suggested harvest of crop residues for off-site bioenergy generation (Paul et al. 1980; Blevins et al. 1983; Elliot and Papendick, 1986). This was spurred by the 1970s energy crisis, and although a portion of the crop residue remaining after grain harvest had traditionally been harvested and used for animal feed or bedding, off-site transport of the crop residues was the critical issue being questioned (Karlen et al., 1984). On-farm use resulted in recycling of nutrients and organic matter via manure disposal, but off-site transport would likely prevent closing field-specific carbon cycles.

## Soil Quality Assessment

Following the 1970s oil crisis, new questions regarding the potential use of crop residues for bioenergy began to emerge as a conservation issue directly linking urban and rural communities. Field research designed to quantify the impact of crop residue harvest on SOM and subsequent productivity resulted in the evolution of a soil health assessment framework. An experiment in southwestern Wisconsin that quantified soil and corn yield response to removing, doubling, or retaining crop residue for 10 years (Karlen et al., 1994a) with moldboard-, chisel-, or no-tillage practices (Karlen et al., 1994b). To more effectively interpret the combined biological, chemical, and physical responses to those treatments, a soil quality/health assessment framework that later becomes known as the Soil Management Assessment Framework (SMAF) (Andrews et al., 2004) was developed. Simultaneously, other assessment tools including an expanded Soil

Conditioning Index (SCI), AgroEcosystem Performance Assessment Tool (AEPAT), and Cornell Soil Health Test (now known as the Comprehensive Assessment of Soil Health or CASH) also began to evolve.

The soil tilth review (Karlen et al., 1990) prompted advancement of soil quality/soil health concepts through a Rodale Foundation workshop (Rodale Institute, 1991) that was described by Haberern (1992) as coming "full circle" in reference to J. I. Rodale's 1942 vision of a "soil-care revolution." Rodale had stated that greater awareness to soil health was needed to create "a healthy society, a country of prosperous farms, and healthy, vigorous people." An important outcome of the Rodale workshop was consensus regarding the need for soil assessments that went beyond productivity and included environmental quality, human and animal health. There was also a realization that assessing and monitoring soil health was complicated by the need to consider multiple soil functions and integrate physical, chemical and biological attributes (Papendick and Parr, 1992; Parr et al., 1992; Warkentin, 1995). Discussions regarding subtle differences between inherent and dynamic soil quality indicators were another important outcome of the Rodale workshop.

Soil quality activities around the world expanded rapidly during the early 1990s, driven in part by increasing recognition of the role soils had in buffering and mitigating factors affecting environmental quality (Warkentin, 1992). However, the true global driver and inspiration for the advancement of soil health or quality was Dr. John W. Doran, to whom this book series is dedicated. His perspective stating that "soil health, or quality, can be broadly defined as the capacity of a living soil to function, within natural or managed ecosystem boundaries, to sustain plant and animal productivity, maintain or enhance water and air quality, and promote plant and animal health" (Doran, 2002) is to us, the ultimate goal for soil health. He also emphasized that soil health will change over time due to natural events or human impacts.

In Australia, Powell and Pratley (1991) developed a "Sustainability Kit" that provided guidelines for measuring soil structure, acidity (pH), salinity, and soil/water temperature. This kit also served as a precursor to John Doran's field-based soil health kit developed in collaboration with scientists at the Rodale Research Center and tested throughout the country (Liebig et al., 1996). Marketed with a USDA Soil Management Manual, the "Soil Health Kit" described simplified tests for soil respiration (Liebig 1996; Doran 1997), infiltration (Ogden et al., 1997), bulk density (Doran 1984), electrical conductivity (EC), pH, and nitrate-nitrogen ($NO_3$-N) concentrations (Smith and Doran, 1996); a soil structure index and penetration test (Bradford and Grossman 1982); soil slaking and aggregate stability tests (Herrick et al. 2001); and an earthworm assessment protocol (Linden 1994). Since that time, the power and potential for on-farm testing of soil health indicators has been greatly magnified by development of the internet, smartphones, and applications such as the Land-Potential Knowledge System (LandPKS; Herrick et al., 2013).

Other developments included a symposium sponsored by the North Central Region Committee No. 59 that focused on SOM at the Soil Science Society of America (SSSA) meetings. That event led to the SSSA publishing two books that became known as the "blue" (Doran et al., 1994) and "green" (Doran and Jones, 1996) soil quality guides that are the precursors for this two-volume series. Doran (2002) also provided important insight stating that although soils have an inherent quality associated with their physical, chemical, and biological properties, their sensitivity to climate and management practices means that land manager decisions ultimately determine soil health. He continued stating a pivotal role for scientists is translating scientific information regarding soil functions into practical tools that land managers can use to evaluate the sustainability of their management practices. Soil health indicators thus became the tools for making the assessments, but there is no single indicator or technology that will always be appropriate.

Soil health assessment has also been advanced by the NRCS, through databases at the Kellogg Soil Survey Laboratory (KSSL), National Soil Survey Center in Lincoln, NE that currently have analytical data for more than 20,000 U.S. pedons and at least 1,100 more from other countries (Brevik et al., 2017). Collectively, morphological descriptions are available for about 15,000 pedons (https://www.nrcs.usda.gov/wps/portal/nrcs/detail/soils/research/?cid=nrcs142p2_053543#:~:text=Summary,1%2C100%20pedons%20from%20other%20countries) (verified 6-12-2020).

The NRCS also has extensive data on basic soil properties, landscape characteristics, and interpretations for use and management. Those databases, describing inherent soil properties, provide a resource to match various land uses with inherent ability of individual soils to perform critical functions (Karlen et al., 2003a). Building upon those resources, the NRCS created several cross-cutting teams during the 1990s, including the Soil Quality Institute (SQI) whose scientists developed many of the first-generation soil quality/soil health scorecards, assessment tools, and information packets. The SQI compiled soil quality information for NRCS staff to help them integrate soil quality concepts into conservation planning and resource inventory activities for their stakeholders (SQI, 1996). The SQI also provided leadership and collaboration for research studies designed to evaluate soil quality indicators at several scales, including Natural Resources Inventory (NRI) sites within four Major Land Resource Areas (Brejda et al., 2000a, 2000b, 2000c).

Linkages between soil and environmental quality were significantly strengthened by the National Academy of Sciences <u>Soil and Water Quality: An Agenda for Agriculture</u> publication (NRC, 1993). It prompted Dr. L.P. Wilding, 1994 president of the SSSA, to appoint a 14-person committee (S-581) with representatives from all SSSA Divisions. Appointees were asked to define the concept of soil quality, examine its rationale, determine if pursuing its development should be a core SSSA activity, and identify soil and plant attributes that would be useful for describing and evaluating soil quality (Karlen et al., 1997).

One of the first comprehensive soil quality/health assessments was used to quantify benefits of the CRP. Following passage of the 1985 Food Security Act, that program took 14.7 million hectares (36.4 million acres) in 36 states out of production (Skold, 1989). Recognized as environmentally sensitive or highly erodible land (HEL), the primary goal was to reduce soil erosion. Secondary CRP goals included: protecting the nation's ability to produce food and fiber, improving air and water quality, carbon sequestration, reducing sedimentation, fostering wildlife habitat, curbing production of surplus commodities, and providing income support for farmers (Allen and Vandever, 2012; FAPRI, 2007; Follett et al., 2001; Li et al., 2017). In exchange for retiring HEL for 10 years, USDA paid CRP participants (farm owners or operators) an annual per-acre rent and half of the cost of establishing a permanent land cover (Young and Osborn, 1990).

Soil quality assessment thus emerged as an evaluation tool as many first-round contracts began to expire (Karlen et al., 1998). Since then, a plethora of studies have evaluated the CRP effect on soil health (Baer et al., 2002; Knops and Tillman, 2000; Matamala et al., 2008; Mensah et al., 2003; Reeder et al., 1998; Rosenzweig et al., 2016; De et al., 2020). Most have focused on insensitive or slower changing soil health indicators (e.g., soil organic carbon [SOC] and N pools). Only a few have evaluated more management-sensitive active carbon pools, such as potentially mineralizable carbon (PMC), potentially mineralizable nitrogen (PMN) microbial biomass carbon (MBC) and soil microbial communities (Baer et al., 2002; Matamala et al., 2008; Haney et al., 2015; Rosenzweig et al., 2016; Li et al., 2018; De et al., 2020).

In Canada, the concept of soil quality/soil health was revived nearly a decade after its introduction by Warkentin and Fletcher (1977) when the Canadian Senate Standing Committee on Agriculture prepared a report on soil degradation (Gregorich, 1996). The research branch of Agriculture and Agri-Food Canada began to work with federal and provincial governments, universities, and the private sector to develop a Soil Quality Evaluation Program (Acton and Gregorich, 1995). Collectiveluy they sought to identify sensitive and reliable indicators (Larson and Pierce, 1991; Haberern 1992; Doran et al., 1994; Doran and Parkin, 1994; Karlen et al., 1994a, 1994b; Karlen et al., 1996). Soil quality assessment began to be interpreted as a sensitive and dynamic way to document soil conditions, response to management, or resistance to the stress imposed by natural forces or human uses (Arshad and Coen, 1992; Haberern, 1992). Their efforts focused on indicator development since soil quality/soil health *per se* cannot be measured directly. This resulted in efforts to identify a minimum set of biological, chemical, and physical measurements (Doran and Parkin, 1996) that provided useful and meaningful information regarding how a specific soil was functioning in response to a specific use and/or management practice (e.g., tillage system, crop rotation, stover harvest, irrigation management, or land-use changes).

In New Zealand, soil quality/soil health assessment was also a rigorously studied topic with one example being "Visual Soil Assessment" (VSA) guidelines for areas characterized as "flat to rolling" or "hill country (Shepherd, 2000; Shepherd et al., 2000a; Shepherd and Janssen, 2000; Shepherd et al., 2000b). Those publications provided instructions, photographs, and scorecards for using VSA to assign scores of 0, 1, or 2 for poor, moderate, or good, respectively, for a variety of soil and plant indicators. Guidelines to help land managers respond if soil quality/soil health was deemed to be either moderate or poor were provided. Inherent site characteristics including land use, soil type, texture, moisture condition, and seasonal weather conditions are also recorded.

Soil quality quickly became closely aligned with good soil management in New Zealand. Furthermore, because land-based industries were the main generator of export income, the use of soil quality assessment as a sustainability indicator generated a substantial amount of research and technology transfer activities throughout that country (Beare et al., 1999). Increasing economic pressure to intensify land use, possibly beyond the margins of sustainability further increased farmer demand for more information, better monitoring tools, and improved soil management practices. Ultimately this led to development of a soil quality monitoring system (SQMS) by Crop & Food Research Ltd. and the Centre for Soil and Environmental Quality (Beare et al., 1999). Several soil quality web sites and other technology transfer activities also evolved during that era.

Many other New Zealand-based soil quality studies that contributed to the foundation upon which current soil health activities have evolved were conducted, but even listing them is beyond the scope of this chapter. One study (Reganold et al., 1993) does warrant discussion because it helps illustrate that soil quality evolution during the 1990s was not without conflict and strong differences in scientific opinion. Reganold et al. (1993) evaluated soil quality and financial performance of biodynamic and conventional farms in New Zealand. They concluded that per unit area, biodynamic farms had better soil quality and were as financially viable as neighboring conventional farms. Within a year a critique questioned the statistical analyses that had been used (Wardle, 1994), but a reanalysis of the data (Reganold, 1994) confirmed the original conclusions and added information indicating that measurements collected from two of the farm pairs had twelve times more earthworms (by number) with biodynamic management than their conventionally managed counterparts.

Bouma (2000) broadened the definition further to include land quality (Bouma et al., 2002) and also explored soil quality effects on the global food supply (Bouma et al., 1998). Meanwhile, Schipper and Sparling (2000) favored the term soil condition, while many German-language publications indicated scientists in that country struggled with differentiating between soil quality and attributes associated with soil fertility (Patzel et al., 2000). Ultimately the German Federal Soil Protection Act [https://germanlawarchive.iuscomp.org/?p=322 (verified 30 June 2020)] was passed to protect or restore soil functions on a permanent sustainable basis. That Act and others in Europe reflected viewpoints that contaminant levels should be the focus for soil quality debates (Singer and Ewing, 2000) and that the concept should be used to improve land use decisions associated with industrial and urban waste and by-product disposal. The goal was to prevent soil property changes that would disrupt natural functions of the soil. Meanwhile, Bujnovskỳ (2000) stressed that for the Slovak Republic, soil quality assessment should focus on deriving a fair soil price and monitoring for degradation. With regard to biomass production, he concluded that was only one of many critical soil functions.

In Australia, soil quality/health research during the 1990s was focused on issues similar to those in New Zealand or the northern hemisphere. For example, Aslam et al. (1999) used MBC, MBN, MBP, and earthworm (*Apporrectodea caligninosa*) populations as biological soil quality indicators to quantify effects of converting pastureland to cropland through either plowing or no-tillage practices. They concluded that conversion using no-till practices could protect soils from biological degradation and maintain better soil quality than with moldboard plowing.

Europe and New Zealand were not the only locations where soil quality was intensively debated. In the United States, Sojka and Upchurch (1999) were fearful SQI and similar efforts could lead to premature conclusions advocating a value system as an end unto itself. They argued there was very little if any parallel between soil, air, and water quality, and that there were regional or taxonomic biases. Karlen et al. (2001) rebutted, emphasizing that advocates and early adopters of soil quality were in total agreement that "our children and grandchildren of 2030 will not care whether we crafted our definitions or diagnostics well. They will care if they are well fed, whether there are still woods to walk in and streams to splash in — in short, whether or not we helped solve their problems, especially given a 30-yr warning." That philosophical debate is mentioned simply to alert soil health advocates the road ahead may not always be smooth.

## 21st Century Developments in Soil Health

In contrast to global exponential growth in soil quality activities during the 1990s, evolution of soil health during the first ten years of the 20th Century showed an identifiable drop (Fig. 2.1). A small contingent of soil scientists continued to question if such holistic research was a good investment. They stated that for more than 200 years soil science had used reductionist research to develop agricultural technologies that unlocked the hidden potential of earth's natural systems to feed, clothe, and provide raw materials to the human population (Sojka et al., 2003). Their perspective was that soil quality proponents sought to change the scientific approach, nomenclature, and institutional priorities for soil science. Proponents countered stating that soil quality was never envisioned as "an end in itself" but rather an evaluation process to identify and quantify soil biological, chemical, and physical responses to specific soil management practices on an identifiable soil resource. They argued that soil health awareness was important because inappropriate, non-sustainable soil management decisions were leading to environmental degradation (e.g., salinization, compaction, erosion, surface and groundwater contamination) and threatening the very soil resources opponents needed to meet 2030 desires of their children and grandchildren. Furthermore, proponents argued soil quality/soil health research, demonstrations, and educational materials simply were evaluating how well a specific soil was functioning in response to both inherent and dynamic properties and processes that can only be evaluated using biological, chemical, and physical indicators. Accusations that proponents of soil health and its assessment was intended to compete, replace, or even diminish the importance of either modern soil survey programs or science-based soil management strategies were totally irrational and unfounded (Karlen et al., 2003b).

The philosophical debates gradually waned and many soil quality proponents quietly moved forward emphasizing soil health which originally had slightly different approaches and priorities than soil quality, but overall were very similar concepts, appropriate for assessing biological production and environmental protection (Harris et al., 1996). An unfortunate consequence of this scientific debate was that public support and even interest in soils began to wane leading to institutional changes that included dismantling the SQI.

Globally, soil health research and technology transfer activities have continued emphasizing:

1) Better soil biological indicators (Stott et al., 2010; Lehman et al., 2015; De et al., 2020)
2) New scoring functions for the SMAF (Wienhold et al., 2009)
3) Commercial availability of CASH (Moebius-Clune et al., 2016)
4) Development of Visual Evaluation of Soil Structure (VESS) techniques (Cherubin et al., 2017; Ball, 2018)
5) Use of least limiting water range (LLWR) to assess soil physical effects (Benjamin and Karlen, 2015)

6) Use of the SMAF or a replacement framework for national (Stott et al., 2011; Karlen et al., 2014b; Veum et al., 2015; Zobeck et al., 2015; Hammac et al., 2016; Ippolito, 2017) and/or international (Fernandez-Ugale et al., 2009; Imaz et al., 2010; Cherubin et al., 2016; Apesteguía et al., 2017) assessments of soil management practices and
7) Use of normalized soil health recovery score (SHRS) to determine which topographic position (e.g., shoulder-, back- , foot- and toe-slope) exhibited the greatest improvement in soil health under CRP (De et al., 2020).

Other developments have included adoption of the Soil Conditioning Index (SCI) by the NRCS as a tool to assess effectiveness of various conservation programs (Soil Quality Institute, 2003). Meanwhile, in Australia the National Soil Research, Development and Extension Strategy identified "soil security" as the foundation for current and future productivity and profitability of agriculture in that country (Koch et al., 2015). Closely paralleling soil health, soil security is to be secured through agricultural land management practices that are matched to the functional capability of a specific soil resource using management practices that improve and maintain soil condition.

Many soil health studies in the United States and other countries during the early 2000s focused on SOM because of its effect on several key biological, chemical, and physical attributes of soil. This included quantifying effects of corn stover harvest (Johnson et al., 2014), biochar amendments (Laird et al., 2010), soil structure and strength characteristics on sloping lands (De et al., 2014), and soil organic carbon (SOC) measurement protocols (Davis et al., 2018). Strategies for placing an economic value on SOM were also explored (Sparling et al., 2006).

An important outcome of continued steady research on soil health indicators was an increased awareness by the private sector that vaulted soil health back into the limelight through programs such as the Soil Renaissance (https://www.farmfoundation.org/projects/the-soil-renaissance-knowledge-to-sustain-earths-most-valuable-asset-1873-d1/ (verified 30 June 2020) and international actions such as the French Government's launch of the "4 per 1000: Soils for Food Security and Climate Initiative" (http://www.regenerationinternational.org/4p1000; verified 30 June 2020).

The Soil Renaissance defined soil health as "the continued capacity of the soil to function as a vital living ecosystem that sustains plants, animals and humans." Their vision, that "improving soil health is the cornerstone of land use management decisions" fostered their mission which was to "reawaken the public to the importance of soil health for enhancing healthy, profitable and sustainable natural resource systems." Meanwhile, the 4 per 1000 declaration sought to: (i) strengthen public policies, tools and actions to support an inclusive and sustainable agricultural development that fosters implementation of farming practices that maintain or enhance carbon in agricultural soils, (ii) mobilize research programs to improve knowledge on soil-carbon storage, assess performance of farming practices and provide science-based evidence to public leaders, and (iii) share our experiences and results, especially science evidence, through a common platform and to regularly organize exchange and stock-taking meetings.

Other developments during the second decade of the 21$^{st}$ Century included formation of the: Soil Health Division (SHD) by the NRCS; Soil Health Institute (SHI) to sustain public/private efforts identified through the Soil Renaissance program; and the Soil Health Partnership (SHP) funded initially by National Corn Growers Association (NCGA), Nature Conservancy (TNC), Monsanto, Walton Family Foundation and more recently the Midwest Row Crop Collaborative, General Mills and many other entities. Without question, the concept of soil health has become wide-spread and has diverse support from a multitude of public and private sectors. Leveraged by this support public and private university soil health programs, non-governmental organization (NGO) soil health activities, a Soil Health Advisory Council associated with the Foundation for Food and Agricultural Research (FFAR), and numerous other groups, soil health has indeed evolved to where it is indeed in the daily vernacular of people around the world.

Perhaps 2500 years after Plato's warning that poor soil management threatens humankind, we may now be ready to embrace soil health as a foundation for healthy landscapes, healthy communities, and to indeed recognize that as Dr. Larson stated many times: Soil is "the thin layer covering the planet that stands between us and starvation"

## References


Acea, M. J., Moore, C. R., and Alexander, M. (1988). Survival and growth of bacteria introduced into soil. *Soil Biology and Biochemistry* 20, 509–515.

Acton, D. F., and Gregorich, L. J. (Eds.). (1995). *The health of our soils – Toward sustainable agriculture in Canada*. Ottawa, Ontario: Agriculture and Agri-Food Canada, Center for Land and Biological Resources Research.

Alexander, M. (1961). *Introduction to soil microbiology*. Hoboken, NJ: John Wiley and Sons.

Alexander, M. (1980). Effects of acidity on microorganisms and microbial processes in soil. In T. C. Hutchinson and M. Havas (Eds.), (pp. 363–374). Boston, MA: Springer.

Allen, A. W., and Vandever, M. W. (2012). *Conservation Reserve Program (CRP) contributions to wildlife habitat, management issues, challenges and policy choices: An annotated bibliography*. USGS Scientific Investigations Report 2012-5066. Reston, VA: USGS.

Allison, F. E. (1968). Soil aggregation: Some facts and fallacies as seen by a microbiologist. *Soil Science* 106, 136–143.

Allison, F. E. (1973). *Soil organic matter and its role in crop production*. New York: Elsevier Scientific Publication Company.

Andrews, S. S., Karlen, D. L., and Cambardella, C. A. (2004). The soil management assessment framework: A quantitative soil quality evaluation method. *Soil Science Society of America Journal* 68, 1945–1962. doi:10.2136/sssaj2004.1945

Apesteguía, M., Virto, I., Orcaray, L., Bescansa, P., Enrique, A., Imaz, M. J., and Karlen, D. L. (2017). Tillage effects on soil quality after three years of irrigation in northern Spain. *Sustainability* 9, 1476–1496. doi:10.3390/su9081476

Arshad, M. A., and Coen, G. M. (1992). Characterization of soil quality: Physical and chemical criteria. *American Journal of Alternative Agriculture* 7, 25–31.

Aslam, T., Choudhary, M. A., and Saggar, S. (1999). Tillage impacts on soil microbial biomass C, N and P, earthworms and agronomy after two years of cropping following permanent pasture in New Zealand. *Soil and Tillage Research* 51, 103–111.

Baer, S. G., Kitchen, D. J., Blair, J. M., and Rice, C.W. (2002). Changes in ecosystem structure and function along a chronosequence of restored grasslands. *Ecological Applications* 12, 1688–1701.

Ball, B. (2018). Visual evaluation of soil structure. Edinburgh: SRUC. http://www.sruc.ac.uk/info/120625/visual_evaluation_of_soil_structure (verified 12 June 2020)

Blevins, R. L., Smith, M. S., Thomas, G. W., and Frye, W. W. (1983). Influence of conservation tillage on soil properties. *Journal of Soil and Water Conservation* 38, 301–305.

Beare, M. H., Williams, P. H., and Cameron, K. C. (1999). On-farm monitoring of soil quality for sustainable crop production. In Currie, L. D., M. J. Hedley, D. J. Horne, and P. Loganathan (Eds.), *Best soil management practices for production* (pp. 81–90). Proceedings of the 1999 Fertilizer and Lime Research Centre Conference, Occasional Report No. 12, Massey Univ., Palmerston North, New Zealand.

Benjamin, J. G., and Karlen, D. L. (2014). LLWR techniques for quantifying potential soil compaction consequences of crop residue removal. *BioEnergy Research* 7, 468–480.

Bennett, H. H. (1950). *Our American Land: The story of its abuse and its conservation. Publication No. 596.* Washington, DC: USDA Soil Conservation Service.

Bouma, J., Batjes, N., and Groot, J. J. H. (1998). Exploring soil quality effects on world food supply. *Geoderma* 86, 43–61.

Bouma, J. (2000). The land quality concept as a means to improve communications about soils. In S. Elmholt, A. Gronlund, V. Nuutinen, and B. Stenberg (Eds.), Proceedings of the NJF Seminar "*Soil Stresses, Quality and Care*". 10-12 April 2000. As Norway, Danish Institute of Agricultural Sciences Reports, Tjele, Denmark.

Bouma, J. (2001). The role of soil science in the land use negotiation process. *Soil Use and Management* 17, 1–6.

Bouma, J. (2002). Land quality indicators of sustainable land management across scales. *Agriculture Ecosystems and Environment* 88, 129–136.

Bradford, J. M., and Grossman, R. B. (1982). In-situ measurement of near-surface soil strength by the fall-cone device. *Soil Science Society of America Journal* 46, 685–688.

Brejda, J. J., Moorman, T. B., Smith, J. L., Karlen, D. L., Allan, D. L., and Dao, T. H. (2000a). Distribution and variability of surface soil properties at a regional scale. *Soil Science Society of America Journal* 64, 974–982.

Brejda, J. J., Moorman, T. M., Karlen, D. L., and Dao, T. H. (2000b). Identification of regional soil quality factors and indicators. I. Central and Southern High Plains. *Soil Science Society of America Journal* 64, 2115–2124.

Brejda, J. J., Karlen, D. L., Smith, J. L., and Allan, D. L. (2000c). Identification of regional soil quality factors and indicators. Part 2. Northern Mississippi loess hills and Palouse prairie. *Soil Science Society of America Journal* 64, 2125–2135.

Brevik, E. C., and Hartemink, A. E. (2010). Early soil knowledge and the birth and development of soil science. *Catena* 83, 23–33.

Brevik, E. C., Pereira, P., Muñoz-Rojas, M., Miller, B. A., Cerdà, A., Parras-Alcántara, L., and Lozano-García, B. (2017). Historical perspectives on soil mapping and process modeling for sustainable land use management. In P. Pereira, E. C. Brevik, M. Muñoz-Rojas, and B. A. Miller (Eds), *Soil mapping and process modeling for sustainable land use management* (pp. 3–28). Netherlands: Elsevier.

Browning, G. M., and Norton, R. A. (1947). Tillage, structure, and irrigation: Tillage practices with corn and soybeans in Iowa. *Soil Science Society of America Proceedings* 12, 491–496.

Bujnovský, R. (2000). Towards the soil quality evaluation. *Ekológia (Bratislava)* 19, 317–323.

Carter, M. C., Gregorich, E. G., Anderson, D. W., Doran, J. W., Janzen, H. H., and Pierce, F. J. (1997). Concepts of soil quality and their significance. In E. G. Gregorich and M. R. Carter (Eds.), *Soil quality for crop production and ecosystem health* (pp. 1–19). Developments in Soil Science 25. Amsterdam, Netherlands: Elsevier Science Publishers B.V.

Cherubin, M. R., Karlen, D. L., Cerri, C. E. P., Franco, A. L. C., Tormena, C. A., Davies, C. A., and Cerri, C. C. (2016). Soil quality indexing strategies for evaluating sugarcane expansion in Brazil. *PLoS ONE* 11, e0150860. doi:10.1371/journal.pone.0150860

Cherubin, M. R., Franco, A. L. C., Guimarães, R. M. L., Tormena, C. A., Cerri, C. E. P., Karlen, D. L., and Cerr, C. C. (2017). Assessing soil structural quality under Brazilian sugarcane expansion areas using Visual Evaluation of Soil Structure (VESS). *Soil and Tillage Research* 175, 64–74.

Davis, M. R., Alves, B. J., Karlen, D. L., Kline, K. L., Galdos, M., and Abulebdeh, D. (2018). Review of soil organic carbon measurement protocols: A U.S. and Brazil comparison and recommendation. *Sustainability* 10, 53. doi:10.3390/su10010053

De, M., Saha, D., and Chakraborty, S. (2014). Soil structure and strength characteristics in relation to slope segments in a degraded Typic Ustroschrepts of North-West India. *Soil Horizons* 55..

De, M., Riopel, J. A., Cihacek, L. J., Lawrinenko, M., Baldwin-Kordick, R., Hall, S. J., and McDaniel, M. D. (2020). Soil health recovery after grassland reestablishment on cropland: The effects of time and topographic position. *Soil Science Society of America Journal* 84, 568–586.

Diamond, J. M. (2005). *Collapse: How societies choose to fail or succeed*. New York: Viking Press.

Doran, J. W. (2002). Soil health and global sustainability: Translating science into practice. *Agriculture, Ecosystems, and Environment* 88, 119–127.

Doran, J. W., Coleman, D. C., Bezdicek, D. F., and Stewart, B. A. (Eds.) (1994). *Defining soil quality for a sustainable environment*. Soil Science Society of America Special Publication No. 35. Madison, WI: SSSA.

Doran, J. W., and Parkin, T. B. (1994). Defining and assessing soil quality. In J. W. Doran et al. (Eds.) *Defining soil quality for a sustainable environment* (pp. 3–21). Soil Science Society of America Special Publication No. 35. Madison, WI: SSSA.

Doran, J. W., and Jones, A. J. (Eds.) (1996). *Methods for assessing soil quality*. Soil Science Society of America Special Publication No. 49. Madison, WI: SSSA.

Doran, J. W., and Parkin, T. B. (1996). Quantitative indicators of soil quality: A minimum data set. In J. W. Doran and A. J. Jones. *Methods for assessing soil quality*. Soil Science Society of America Special Publication No. 49. Madison, WI: SSSA.

Doran, J., Kettler, T., and Tsivou, M. (1997). *Field and laboratory Solvita soil test evaluation*. Lincoln, NE: University of Nebraska USDA-ARS.

Dormaar J. F., Lindwall, C. W., and Kozub, G. C. (1988). Effectiveness of manure and commercial fertilizer in restoring productivity of an artificially eroded Dark Brown Chernozemic soil under dryland conditions. *Canadian Journal of Soil Science* 68, 669–679.

Elliott, L. F, and Papendick, R. I. (1986). Crop residue management for improved soil productivity. *Biological Agriculture and Horticulture* 3, 131–142.

FAPRI. (2007). *Estimating water quality, air quality, and soil carbon benefits of the Conservation Reserve Program*. Columbia, MO: FAPRI.

Faulkner, E. H. (1943). Plowman's folly. Reprint Uni. of Oklahoma Press, 2012.

Fernandez-Ugale, O., Virto, I., Bescansa, P., Imaz, M. J., Enrique, A., and Karlen, D. L. (2009). No-tillage improvement of soil physical quality in calcareous, degradation-prone, semiarid soils. *Soil & Tillage Research* 106, 29–35.

Follett, R. F., Pruessner, E. G., Samson-Liebig, S. E., Kimble, J. M., and Waltman, S.W. (2001). Carbon sequestration under the Conservation Reserve Programin the historic grassland soils of the United States of America. In R. Lal (Ed.), *Carbon sequestration and the greenhouse effect* (pp. 27–40). Madison, WI: SSSA.

Fream, W. (1890). Tilth. In *Soils and their properties* (pp. 95–100). London: George Bell & Sons.

Ghilarov, M. S. (1983). Darwin's formation of vegetable mould: Its philosophical basis. In *Earthworm ecology* (pp. 1–4). Dordrecht: Springer.

Gregorich, E. G. (1996). Soil quality: A Canadian perspective. Proceedings of the Soil Quality Indicators Workshop, 8-9 Feb. 1996, *Ministry of Agriculture and Fisheries, and Lincoln Soil Quality Research Center*. Christchurch, NZ: Lincoln University.

Haberern, J. (1992). Coming full circle – The new emphasis on soil quality. *American Journal of Alternative Agriculture* 7, 3–4.

Hammac, W. A., Stott, D. E., Karlen, D. L., and Cambardella, C. A. (2016). Crop, tillage, and landscape effects on near-surface soil quality indices in Indiana. *Soil Science Society of America Journal* 80, 1638–1652.

Haney, R. L., Haney, E. B., Hossner, L. R., and Arnold, J. G. (2006). A new soil extraction for simultaneous phosphorus ammonium, and nitrate analysis. *Communications in Soil Science and Plant Analysis* 37, 1511–1523.

Haney, R. L., Haney, E. B., Smith, D. R., and White, M. J. (2015). Estimating potential nitrogen mineralisation using the Solvita soil respiration system. *Open Journal of Soil Science* 5, 319–323. doi.org/10.4236/ojss.2015.512030

Harris, R. F., Karlen, D. L., and Mulla, D. (1996). A conceptual framework for assessment and management of soil quality and health. In J. W. Doran, D. C. Coleman, D. F. Bedzicek and B. A. Stewart (Eds.), *Methods for assessing soil quality* (pp. 61–82). Soil Science Society of America Special Publication No. 49. Madison,WI: SSSA.

Herrick, J. E., Whitford, W.G., De Soyza, A.G., Van Zee, J. W., Havstad, K. M., Seybold, C. A., and Walton, M. (2001). Field soil aggregate stability kit for soil quality and rangeland health evaluations. *Catena* 44, 27–35.

Herrick, J. E., Urama, K. C., Karl, J. W., Boos, J., Johnson, M. V. V., Shepherd, K. D., Hempel, J., Bestelmeyer, B. T., Davies, J., Guerra, J. L., Kosnik, C., Kimiti, D. W., Ekai, A. L., Muller, K., Norfleet, L., Ozor, N., Reinsch, T., Sarukhan, J., and West, L. T. (2013). The global Land-Potential Knowledge System (LandPKS): Supporting evidence-based, site-specific land use and management through cloud computing, mobile applications, and crowdsourcing. *Journal of Soil and Water Conservation* 68, 5A–12A. doi:10.2489/jswc.68.1.5A

Hillel, D. (1991). *Out of the earth: Civilization and the life of the soil*. Oakland, CA: University of California Press.

History.com Editors. (2020). Dust Bowl. https://www.history.com/topics/great-depression/dust-bowl (Verified 30 June 2020).

Imaz, M. J., Virto, I., Bescansa, P., Enrique, A., Fernandez-Ugalde, O., and Karlen, D. L. (2010). Soil quality indicator response to tillage and residue management on semi-arid Mediterranean cropland. *Soil & Tillage Research* 107, 17–25.

Ippolito, J. A., Bjorneberg, D., Stott, D., and Karlen, D. (2017). Soil quality improvement through conversion to sprinkler irrigation. *Soil Science Society of America Journal* 81, 1505–1516. doi:10.2136/sssaj2017.03.0082

Jenny, H. (1980). *The soil resource*. New York: Springer-Verlag.

Johnson, D. L., and Schaetzl, R. J. (2015). Differing views of soil and pedogenesis by two masters: Darwin and Dokuchaev. *Geoderma* 237, 176–189.

Johnson, J. M. F., Novak, J. M., Varvel, G. E., Stott, D. E., Osborne, S. L., Karlen, D. L., Lamb, J. A., Baker, J., and Adler, P. R. (2014). Crop residue mass needed to maintain soil organic carbon levels: Can it be determined? *BioEnergy Research* 7, 481–490.

Karlen, D. L., Hunt, P. G., and Campbell, R. B. (1984). Crop residue removal effects on com yield and fertility of a Norfolk sandy loam. *Soil Science Society of America Journal* 48, 868–872.

Karlen, D. L., Erbach, D. C., Kaspar, T. C., Colvin, T. S., Berry, E. C., and Timmons, D. R. (1990). Soil tilth: A review of past perceptions and future needs. *Soil Science Society of America Journal* 54, 153–161.

Karlen, D. L., Wollenhaupt, N. C., Erbach, D. C., Berry, E. C., Swan, J. B., Eash, N. S., and Jordahl, J. L. (1994a). Crop residue effects on soil quality following 10-years of no-till corn. *Soil and Tillage Research* 31, 149–167.

Karlen, D. L., Wollenhaupt, N. C., Erbach, D. C., Berry, E. C., Swan, J. B., Eash, N. S., and Jordahl, J. L. (1994b). Long-term tillage effects on soil quality. *Soil and Tillage Research* 32, 313–327.

Karlen, D. L., Parkin, T. B., and Eash, N. S. (1996). Use of soil quality indicators to evaluate Conservation Reserve Program sites in Iowa. In J. W. Doran and A. J. Jones (Eds.) *Methods for assessing soil quality* (pp. 345–355). Soil Science Society of America Special Publication No. 49. Madison, WI: SSSA.

Karlen, D. L., Mausbach, M. J., Doran, J. W., Cline, R. G., Harris, R. F., and Schuman, G. E. (1997). Soil quality: A concept, definition, and framework for evaluation. *Soil Science Society America Journal* 61, 4–10.

Karlen, D. L., Gardner, J. C., and Rosek, M. J. (1998). A soil quality framework for evaluating the impact of CRP. *Journal Production Agriculture* 11, 56–60.

Karlen, D. L., Andrews, S. S., and Doran, J. W. (2001). Soil quality: Current concepts and applications. *Advances in Agronomy* 74, 1–40.

Karlen, D. L., Ditzler, C. A., and Andrews, S. S. (2003a). Soil quality: Why and how? *Geoderma* 114, 145–156.

Karlen, D. L., Andrews, S. S., Wienhold, B. J., and Doran, J. W. (2003b). Soil quality: Humankind's foundation for survival. *Journal of Soil Water Conservation* 58, 171–179.

Karlen, D. L., Peterson, G. A., and Westfall, D. G. (2014a). Soil and water conservation: Our history and future challenges. *Soil Science Society of America Journal* 78, 1493–1499. doi:10.2136/sssaj2014.03.0110

Karlen, D. L., Stott, D. E., Cambardella, C. A., Kremer, R. J., King, K. W., and McCarty, G. W. (2014b). Surface soil quality in five Midwest cropland CEAP watersheds. *Journal Soil Water Conservation* 69, 393–401.

Karlen, D. L., and Rice, C. W. (2015). Soil degradation: Will humankind ever learn? *Sustainability* 7, 12490–12501. doi:10.3390/su70912490

Keen, B. A. (1931). *The physical properties of the soil.* Rothamsted monograph on Agricultural Science. London: Ser. Longmans, Green and Co.

Klingebiel, A. A., and O'Neal, A. M. (1952). Structure and its influence on tilth of soils. *Soil Science Society of America Journal* 16, 77–80.

Knops, J. M. H., and Tilman, D. (2000). Dynamics of soil nitrogen and carbon accumulation for 61 years after agricultural abandonment. *Ecology* 81, 88–98.

Koch, A., Chappell, A., Eyres, M., and Scott, E. (2015). Monitor soil degradation or triage for soil security? An Australian challenge. *Sustainability* 7, 4870–4892. doi:10.3390/su7054870

Laird, D. A., Fleming, P. D., Davis, D. D., Horton, R., Wang, B., and Karlen, D. L. (2010). Impact of biochar amendments on the quality of a typical Midwestern agricultural soil. *Geoderma* 158, 443–449.

Larson, W. E., and Pierce, F. J. (1991). Conservation and enhancement of soil quality. In *Evaluation for sustainable land management in the developing world* (pp. 175–203). Vol 2. IBSRAM Proc. 12(2). Bangkok, Thailand: International Board for Soil Research and Management.

Lehman, R. M., Cambardella, C. A., Stott, D. E., Acosta-Martinez, V., Manter, D. K., Buyer, J. S., Maul, J. E., Smith, J. L., Collins, H. P., Halvorson, J. J., Kremer, R. J., Lundgren, J. G., Ducey, T. F., Jin, V. L., and Karlen, D. L. (2015). Understanding and enhancing soil biological health: The solution for reversing soil degradation. *Sustainability* 7, 988–1027.

Li, C., Fultz, L. M., Moore-Kucera, J., Acosta-Martínez, V., Horita, J., Strauss, R., Zak, J., Calderóng, F., and Weindorf, D. (2017). Soil carbon sequestration potential in semi-arid grasslands in the Conservation Reserve Program. *Geoderma* 294, 80–90.

Li, C., Fultz, L. M., Moore-Kucera, J., Acosta-Martínez, V., Kakarla, M., and Weindorf, J. D. (2018). Soil microbial community restoration in Conservation Reserve Program semi-arid grasslands. *Soil Biology and Biochemistry* 118, 166–177. https://doi.orrg/10,1016/j.soilbio.2017.12,001

Liebig, M. A., Doran, J. W., and Gardner, J. C. (1996). Evaluation of a field test kit for measuring selected soil quality indicators. *Agronomy Journal* 88, 683–686.

Linden, D. R., Hendrix, P. F., Coleman, D. C., and van Vliet, P. C. (1994). Faunal indicators of soil quality. In J. W. Doran, D. C. Coleman, D. F. Bezdicek, and

B. A. Stewart (Eds.), *Defining soil quality for a sustainable environment* (pp. 91–106). SSSA Special Publication No. 35. Madison, WI: SSSA.

Lowdermilk, W. C. (1953). *Conquest of the land through 7000 years.* SCS Agriculture Bulletin No. 99. Washington, DC: USDA Soil Conservation Service.

Lyon, T. L., Buckman, H. O., and Brady, N. C. (1950). *The nature and properties of soils.* 5th ed. New York: MacMillan Co.

Martin, J. P., and Waksman, S. A. (1939). The role of microorganisms in the conservation of the soil. *Science* 90, 304–305.

Martin, J. P., and Waksman, S. A. (1941). Influence of microorganisms on soil aggregation and erosion: II. *Soil Science* 52, 381–394.

Martin, J. P. (1971). Decomposition and binding action of polysaccharides in soil. *Soil Biology and Biochemistry* 3, 33–41.

Matamala, R., Jastrow, J. D., Miller, R. M., and Garten, C. T. (2008). Temporal changes in C and N stocks of restored prairie: Implications for C sequestration strategies. *Ecological Applications* 18, 1470–1488. https://doi.org/10.1890/07-1609.1

Mathur, S. P., and Paul, E. A. (1967). Microbial utilization of soil humic acids. *Canadian Journal of Microbiology* 13, 573–580.

Melsted, S. W. (1954). New concepts of management of Corn Belt soils. *Advances in Agronomy* 6, 121–142.

Mensah, F., Schoenau, J. J., and Malhi, S. S. (2003). Soil carbon changes in cultivated and excavated land converted to grasses in east-central Saskatchewan. *Biogeochemistry* 63, 85–92. https://doi.org/10.1023/A:1023369500529

Michel, J. B., Shen, Y. K., Aiden, A. P., Veres, A., Gray, M. K., Pickett, J. P., Hoiberg, D., Clancy, D., Norvig, P., Orwant, J., Pinker, S., Nowak, M. A., and Aiden, E. L. (2011). Quantitative analysis of culture using millions of digitized books. *Science* 331, 176–182.

Moebius-Clune, B. N., Moebius-Clune, D. J., Gugino, B. K., Idowu, O. J., Schindelbeck, R. R., Ristow, A. J., van Es, H. M., Thies, J. E., Shayler, H. A., McBride, M. B., Kurtz, K. S., Wolfe, D. W., and Abawi, G. S. (2016). *Comprehensive assessment of soil health – The Cornell Framework.* Edition 3.2. Geneva, NY: Cornell University.

Montgomery, D. R. (2007). *Dirt: The erosion of civilizations.* Berkley, CA: University of California Press.

National Research Council (NRC). (1993). *Soil and water quality: An agenda for agriculture.* Washington, DC: National Academy Press.

NRCS. (2018). More than 80 years helping people help the land: A brief history of NRCS. https://www.nrcs.usda.gov/wps/portal/nrcs/detail/national/about/history/?cid=nrcs143_021392 (verified 30 June 2020).

Papendick, R. I., and Parr, J. F. (1992). Soil quality – The key to a sustainable agriculture. *American Journal of Alternative Agriculture* 7, 2–3.

Parr, J. F., Papendick, R. I., Hornick, S. B., and Meyer, R. E. (1992). Soil quality: Attributes and relationship to alternative and sustainable agriculture. *American Journal of Alternative Agriculture* 7, 5–11.

Patzel, N., Sticher, H., and Karlen, D. (2000). Soil fertility – Phenomenon and concept. *Journal of Plant Nutrition and Soil Science* 163, 129–142.

Paul, E. A., and Voroney, R. P. (1980). Nutrient and energy flows through soil microbial biomass. In D. C. Ellwood, J. N. Hedger, M. J. Latham, J. M. Lynch, and J. H. Slater (Eds.), *Contemporary microbial ecology* (pp. 215–237). Proceedings of the Second International Symposium on Microbial Ecology, 7-12 Sept. 1980, University of Warwick, Coventry, UK. Academic Press Inc.

Paul, E., Rasmussen, R. R., Allmaras, C. R., Rohde, C. R., and Roager, N. C., Jr. (1980). Crop residue influences on soil carbon and nitrogen in a wheat-fallow system. *Soil Science Society of America Journal* 44, 596–600.

Paul, E. A. (2014). *Soil microbiology, ecology and biochemistry*. Academic Press.

Pawluk, R. R., Sandor, J. A., and Tabor, J. A. (1992). The role of indigenous soil knowledge in agricultural development. *Journal of Soil and Water Conservation* 47, 298–302.

Pierce, F. J., Larson, W. E., Dowdy, R. H., and Graham, W.A.P. (1983). Productivity of soils: Assessing long-term changes due to erosion. *Journal of Soil and Water Conservation* 38, 39–44

Pierce, F. J., Larson, W. E., Dowdy, R. H., and Graham, W. A. P. (1984). Soil productivity in the Corn Belt: An assessment of erosion's long-term effects. *Journal of Soil and Water Conservation* 39, 131–136.

Powell, D., and Pratley, J. (1991). *Sustainability kit manual. Centre for Conservation Farming*. Wagga Wagga, Australia: Charles Sturt University-Riverina.

Raji, B. A., Malgwi, W. B., Chude, V. O., and Berding, F. (2006). Integrating indigenous knowledge and conventional soil science approaches to detailed soil survey in Kaduna State, Nigeria. In Proceedings of the 18$^{th}$ World Congress of Soil Science, July 2006, Philadelphia, PA.

Reeder, J. D., Schuman, G. E., and Bowman, R. A. (1998). Soil C and N changes on conservation reserve program lands in the Central Great Plains. *Soil and Tillage Research* 47, 339–349. https://doi.org/10.1016/S0167-1987(98)00122-6

Reganold, J. P., Palmer, A. S., Lockhart, J. C., and Macgregor, A. N. (1993). Soil quality and financial performance of biodynamic and conventional farms in New Zealand. *Science* 260, 344–349.

Reganold, J. P. (1994). Response to statistical analyses of soil quality. *Science* 264, 282–283.

Rodale Institute. (1991). Conference report and abstracts. International conference on the Assessment and Monitoring of Soil Quality. Emmaus, PA, 11-14 July 1991. Emmaus, PA: Rodale Press.

Rosenzweig, S. T., Carson, M. A., Baer, S. G., and Blair, J. M. (2016). Changes in soil properties, microbial biomass, and fluxes of C and N in soil following post-agricultural grassland restoration. *Applied Soil Ecology* 100, 186–194. https://doi.org/10.1016/j.apsoil.2016.01.001

Schipper, L. A., and Sparling, G. P. (2000). Performance of soil condition indicators across taxonomic groups and land uses. *Soil Science Society of America Journal* 64, 300–311.

Shepherd, T. G. (2000). *Visual soil assessment. Volume 1. Field guide for cropping and pastoral grazing on flat to rolling country*. Palmerston North, New Zealand: Horizons.mw/Landcare Research.

Shepherd, T. G., Ross, C. W., Basher, L. R., and Saggar, S. (2000a). *Visual soil assessment. Volume 2. Soil management guidelines for cropping and pastoral grazing on flat to rolling country*. Palmerston North, New Zealand: Horizons.mw/Landcare Research.

Shepherd, T. G., and Janssen, H. J. (2000). *Visual soil assessment. Volume 3. Field guide for hill country land uses*. Palmerston North, New Zealand: Horizons.mw/Landcare Research.

Shepherd, T. G., Janssen, H. J., and Bird, L. J. (2000b). *Visual soil assessment. Volume 4. Soil management guidelines for hill country land uses*. Palmerston North, New Zealand: Horizons.mw/Landcare Research.

Singer, M. J., and Ewing, S. (2000). Soil quality. In M. E. Sumner (Ed.), *Handbook of soil science* (pp. G-271–G-298). Boca Raton, FL: CRC Press.

Skold, M. D. (1989). Cropland retirement policies and their effects on land use in the Great Plains. *Journal of Production Agriculture* 2, 197–201.

Smith, J. L., and Doran, J. W. (1996). Measurement and use of pH and electrical conductivity for soil quality analysis. In J. W. Doran and A. J. Jones (Eds.) *Methods for assessing soil quality* (pp. 169–185). Soil Science Society of America Special Publication No. 49. Madison, WI: SSSA.

Soil Quality Institute (SQI). (1996). The soil quality concept. Edited by Soil Quality Institute. USDA-NRCS.

Soil Quality Institute (SQI). (2003). *Interpreting the soil conditioning index: A tool for measuring soil organic matter trends*. Soil Quality – Agronomy Technical Note No. 16. Auburn, AL: Soil Quality Institute. https://www.nrcs.usda.gov/Internet/FSE_DOCUMENTS/nrcs142p2_053273.pdf (verified 6-12-2020)

Sojka, R. E., and Upchurch, D. R. (1999). Reservations regarding the soil quality concept. *Soil Science Society of America Journal* 63, 1039–1054.

Sojka, R. E., Upchurch, D. R., and Borlaug, N. E. (2003). Quality soil management or soil quality management: Performance vs semantics. *Advances in Agronomy* 79, 1–68.

Sparling, G. P., Wheeler, D., Vesely, E.-T., and Schipper, L. A. (2006). What is soil organic matter worth? *Journal of Environmental Quality* 35, 548–557.

Stott, D. E., Andrews, S. S., Liebig, M. A., Wienhold, B. J., and Karlen, D. L. (2010). Evaluation of β-glucosidase activity as a soil quality indicator for the Soil Management Assessment Framework (SMAF). *Soil Science Society America Journal* 74, 107–119.

Stott, D. E., Cambardella, C. A., Wolf, R., Tomer, M. D., and Karlen, D. L. (2011). A soil quality assessment within the Iowa River south fork watershed. *Soil Science Society America Journal* 75, 2271–2282.

Van Bavel, C. H. M., and Schaller, F. W. (1950). Soil aggregation, organic matter and yields in a long-time experiment as affected by crop management. *Soil Science Society of America Proceedings* 15, 399–408.

Van Doren, C. A., and Klingebiel, A. A. (1952). Effect of management on soil permeability. *Soil Science Society of America Journal* 16, 66–69.

Veum, K. S., Kremer, R. J., Sudduth, K. A., Kitchen, N. R., Lerch, R. N., Baffaut, C., Stott, D. E., Karlen, D. L., and Sadler, E. J. (2015). Conservation effects on soil quality indicators in the Missouri Salt River Basin. *Journal of Soil and Water Conservation* 70, 232–246.

Voorhees, W. B., Senst, C. G., and Nelson, W. W. (1978). Compaction and soil structure modification by wheel traffic in the northern Com Belt. *Soil Science Society of America Journal* 42, 344–349.

Voorhees, W. B. (1979). Soil tilth deterioration under row cropping in the northern Corn Belt: Influence of tillage and wheel traffic. *Journal of Soil and Water Conservation* 34, 184–186.

Voorhees, W. B. (1983). Relative effectiveness of tillage and natural forces in alleviating wheel-induced soil compaction. *Soil Science Society of America Journal* 47, 129–133.

Voorhees, W. B., and Lindstrom, M. J. (1984). Long-term effects of tillage method on soil tilth independent of wheel traffic compaction. *Soil Science Society of America Journal* 48, 152–156.

Waksman, S. A., and Starkey, R. L. (1924). Microbiological analysis of soil as an index of soil fertility: VII. Carbon dioxide evolution. *Soil Science* 17, 141–162.

Waksman, S. A., and Hutchings, I. J. (1935). The role of plant constituents in the preservation of nitrogen in the soil. *Soil Science* 40, 487–497

Wardle, D. A. (1994). Statistical analyses of soil quality. *Science* 264, 281–282.

Warkentin, B. P., and Fletcher, H. F. (1977). Soil quality for intensive agriculture. In *Proceedings of the International Seminar on Soil Environment and Fertilizer Management in Intensive Agriculture* (pp. 594–598). Japan: Society of Science of Soil and Manure.

Warkentin, B. P. (1992). Soil science for environmental quality – How do we know what we know? *Journal of Environmental Quality* 21, 163–166.

Warkentin, B. P. (1995). The changing concept of soil quality. *Journal of Soil and Water Conservation* 50 226–228.

Warkentin, B. P. (2008). Soil structure: A history from tilth to habitat. *Advances in Agronomy* 97, 239–272.

Whiteside, E. P., and Smith, R. S. (1941). Soil changes associated with tillage and cropping in humid areas of the United States. *Agronomy Journal* 33, 765–777.

Wienhold, B. J., Karlen, D. L., Andrews, S. S., and Stott, D. E. (2009). Protocol for indicator scoring in the Soil Management Assessment Framework (SMAF). *Renewable Agriculture Food Systems* 24, 260–266.

Wilson, H. A., and Browning, G. M. (1945). Soil aggregation, yields, runoff and erosion as affected by cropping systems. *Soil Science Society of America Proceedings* 10, 51–57.

Yoder, R. E. (1937). The significance of soil structure in relation to the tilth problem. *Soil Science Society of America Proceedings* 2, 21–33.

Young, C. E., and Osborn, C. T. (1990). Costs and benefits of the conservation reserve program. *Journal of Soil and Water Conservation* 45, 370–373.

Zobeck, T. M., Steiner, J. L., Stott, D. E., Duke, S. E., Starks, P. J., Moriasi, D. N., and Karlen, D. L. (2015). Soil quality index comparisons using Fort Cobb, Oklahoma, watershed-scale land management data. *Soil Science Society of America Journal* 79, 224–238.

# 3

# The Utility and Futility of Soil Health Assessment

*John F. Obrycki and Lumarie Pérez-Guzmán*

## Chapter Overview

Documenting benefits from soil health management practices and assessments has been described as both useful and futile because it requires continual observation, some form of data collection, and an assessment protocol. This chapter focuses on the benefits of soil health being evaluated through soil physical, chemical, and biological property measurements. A producer, landowner, or researcher interested in soil health usually wants to know if soil properties are changing from an identifiable condition or point of interest, such as an inherent baseline or an equilibrium condition established by business-as-usual soil and crop management practices. When soils are considered within social, political, economic, and environmental contexts, the type of benefits that can be documented expands (Heller and Keoleian, 2003; McBratney et al., 2014; Mena Mesa et al., 2014; Rasul and Thapa, 2004; Steffan et al., 2017; Wolde et al., 2016), but although those assessment scales are important to consider, they are outside the scope of this chapter because such changes, whether positive or negative, generally take several years (perhaps even decades) to be noticeable and/or measurable. This chapter focuses on agricultural research and discusses the general opportunities and limitations associated with soil health management approaches and strategies used to document potential soil physical, chemical, and biological property changes.

*Soil Health Series: Volume 1 Approaches to Soil Health Analysis*, First Edition.
Edited by Douglas L. Karlen, Diane E. Stott, and Maysoon M. Mikha.

## Introduction

There are several important questions associated with soil health research (Fig. 3.1). These include issues associated with more clearly defining the soil health concept, determining how to measure and quantify soil health at multiple scales, and using these principles to guide current and future soil and crop management decisions. As discussed in Chapter 2, questions regarding how to achieve effective soil management are not new (e.g., Bennett and Chapline, 1928; Hobbs, 2007; Janvier et al., 2007; Janzen, 2001; Karlen et al., 1997; Karlen et al., 2019; Magdoff and van Es, 2009; Stoll, 2003). Furthermore, several visual, in-field, and laboratory methods for evaluating soil health have been developed over several decades. Answers to those questions are not simple because the living and dynamic nature of soils results in fiscal, human resource, intellectual, and other research constraints associated with sampling, analyzing, and interpreting how soil biological, chemical, and physical properties and processes affect soil health. Our objectives for this chapter are to provide practical definitions and examples of various approaches for addressing soil health, along with our assessment of current analytical methods, their limitations, and potential research topics that may clarify and help advance the concept.

### Definitions

Within the context of soil health, we suggest the term "benefit" refers to a human defined and desired change in soil physical, chemical, and/or biological

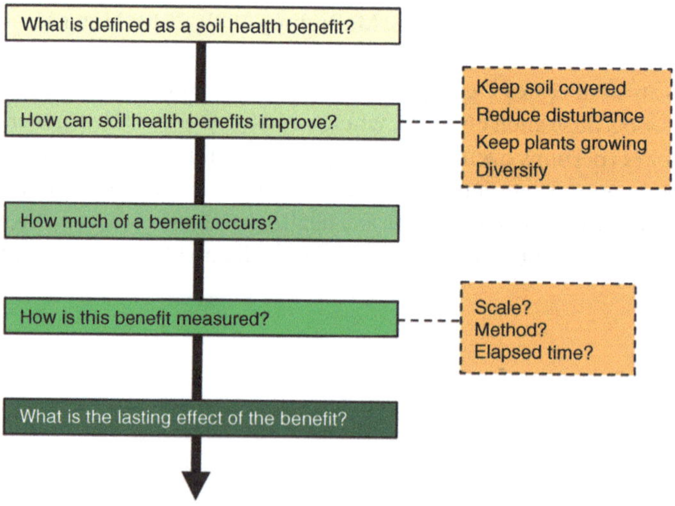


**Figure 3.1** Key soil health research questions and selected responses (orange boxes).

properties and processes because of their effect on critical functions (*i.e.*, productivity, filtering and buffering, water entry, retention and release) that soils provide for humankind. For example, a decrease in soil compaction, and an associated increase in soil porosity, would be considered a physical soil health benefit because less compacted soils allow more water to infiltrate and be stored in the soil. Those water and retention benefits subsequently improve productivity as plants are able to extract and use the soil water, and provide environmental benefits because the rate and amount of water running off the site (carrying soil particles, nutrients, pesticides, etc.) is decreased. Soil pH can be used to illustrate the role of chemical properties and processes in a soil health assessment because it affects critical soil functions such as nutrient availability and fitness for plant root growth and development. With regard to soil biological properties and processes, soil organic matter (SOM) or a closely associated measurement (*i.e.*, active carbon, β-glucosidase, particulate organic matter) can be monitored over time to determine how various soil and crop management practices are affecting the soil (Gebhart et al., 1994; Li et al., 2017). In general, it would be considered a benefit if SOM increases because it can then increase soil water holding capacity, nutrient retention and cycling, and soil aggregation.

A second important point when defining soil health benefits is to realize that because of the living and dynamic nature of soils, changes are site- and landscape-specific and therefore when interpreting the relative importance of a change, the phrase "it depends" must be kept in mind. For each soil biological, chemical, and physical soil health indicator, there are ranges over which changes are of most interest and highly influential as well as other ranges where they have minimal to no agronomic, environmental, or other economically important effect.

Previously, soil health indicator benefits have generally been conceptualized as following one of three curve types: "less is better" (e.g., soil compaction), "more is better" (e.g., SOM content), and "mid-point optimum" (e.g., soil pH) (Andrews et al., 2004; Moebius-Clune et al., 2016). Therefore, a soil with 500 g $kg^{-1}$ (50%) organic matter may be a suitable peat or wetland soil with environmental buffering, wildlife, or other positive attributes, but without major investment in drainage water management, it would not be a suitable soil for production of corn (*Zea mays* L.), soybean (*Glycine max* [L.] Merr.), wheat (*Triticum aestivum* L.) or cotton (*Gossypium hirsutum* L.). Similarly, an acidic soil is desired for high-bush blueberries (*Vaccinium corymbosum* L.) or some forest species, but would be toxic for plants that cannot tolerate the high concentrations of soluble aluminum (Al) or manganese (Mn) that can occur under those conditions. Soil health, therefore, does not mean that all soils will have the same properties, but all soils will exhibit health benefits when physical, chemical, and biological properties are evaluated in the context of one or more specific soil functions.

The third and final focus that needs to be defined for this chapter is the phrase "soil health approaches". We use this term to refer to management systems that consider soil physical, chemical, and biological properties collectively, rather than focusing on only one aspect (Andrews et al., 2004; Moebius-Clune et al., 2016). For example, adequate nutrient availability for plants provided by routine soil fertility testing and good fertilizer management is not sufficient for a soil to be considered "healthy" if that resource is highly compacted due to excessive or inappropriate wheel traffic, eroded by wind or water, or depleted in SOM compared to its inherent conditions. Soil health approaches must focus on comprehensive management that views soil resources as physical, chemical, and biological systems and uses practices that address all three components. Implementation of such approaches is not difficult and may be accomplished by combining routine soil test recommendations with reduced tillage intensities, controlled traffic planting, and harvest patterns. Collectively, such a soil health approach could improve soil nutrient availability and reduce soil compaction.

**Opportunities for Implementing Soil Health Approaches**

The primary purpose for developing and implementing soil health approaches is to encourage the use of scientifically-based, comprehensive soil management practices that account for not only economic goals such as productivity but also how the entire landscape is maintained (Schnepf and Cox, 2006). Although soil health approaches per se have become recognized research and technology transfer topics during the past two decades, to many readers the concept is not new but rather another term transferable to previous efforts defined as soil conservation, soil quality, soil condition, soil tilth, or simply soil management. For example, Table 3.1 lists four soil and plant management strategies often presented as 21st century soil health approaches that have been considered to be good soil management practices throughout not only the 20th century (King, 1911; Keen, 1931; Lowdermilk, 1953; Keen, 1931) but from the time of Plato and before (Fream, 1890; Hillel, 1991; Diamond, 2011; Montgomery, 2007). Without question, those four management practices are in no way comprehensive, but they do provide a general framework under which many soil health approaches can be listed. A common link between the practices is that they strive to improve soils by keeping them covered and protected by erosive forces of wind or water and they strive to keep soil biological and chemical cycles as active as possible (Table 3.2). Keeping soils covered with living plants minimizes runoff, encourages proliferation of roots, production of root exudates and recycling of senesced vegetation as food for soil microbes, and provides a biotic pump for moving water through the soil–plant–atmosphere continuum. Operations, including planting date and rates, field equipment, traffic patterns, crop sequences,

**Table 3.1** Timeless generic strategies for improving soils.

| | Year | | |
|---|---|---|---|
| **Management Practice** | **1937[a]** | **2017[b]** | **2097[c]** |
| Keep soil covered | ✓ | ✓ | ✓ |
| Reduce soil disturbance | ✓ | ✓ | ✓ |
| Keep plants growing year round | ✓ | ✓ | ✓ |
| Diversify | ✓ | ✓ | ✓ |

[a] Adapted from Rule, 1937.
[b] Taken from guidelines presented by the USDA-Natural Resources Conservation Service (NRCS), 2019b.
[c] Our projection of soil health approaches that will continue to be emphasized.

With regard to 21st century soil health approaches, two questions emerging from the generic guidelines in Table 3.1 are: (1) What methods can be used to implement practices that will effectively achieve the underlying management goals?, and (2) What magnitude of soil health benefit can be achieved by investing in these practices? To help address the first question, several USDA-NRCS Conservation Practice Standards (Table 3.2) have been developed (Schnepf and Cox, 2006; USDA-NRCS, 2019a). It is important to note, however, that each "practice standard" can apply to multiple conservation management options that have multiple interactions as illustrated in (Fig. 3.2). This uncertainty is also one reason soil health is considered useful by some and futile by others.

**Table 3.2** Selected NRCS conservation practices identified as also having a soil health impact.

| **Conservation Goal** | **Potential NRCS Approved Practices** |
|---|---|
| Keep soil covered | Conservation cover (327[a]); Forage and biomass planting (512); Mulching (484) |
| Reduce soil disturbance | Contour farming (330); Controlled traffic farming (334); Residue and tillage management (329 and 345); Strip cropping (585); Windbreak (380) |
| Keep plants growing year round | Conservation crop rotation (328); Cover crop (340); Wildlife habitat management (645) |
| Diversify | Contour buffer strips (332); Filter strip (393); Grassed waterway (412); Riparian forest buffer (391) |

[a] Specific Conservation Practice Numbers associated with practices approved by USDA-NRCS (2019a).

When considering the conservation practices listed in Table 3.2 or elsewhere as strategies to remediate or enhance soil health, it is very important to recognize that implementation will require producer dedication, more time, and perhaps greater financial investment than most business as usual operations. Producers may need to change several core components of their

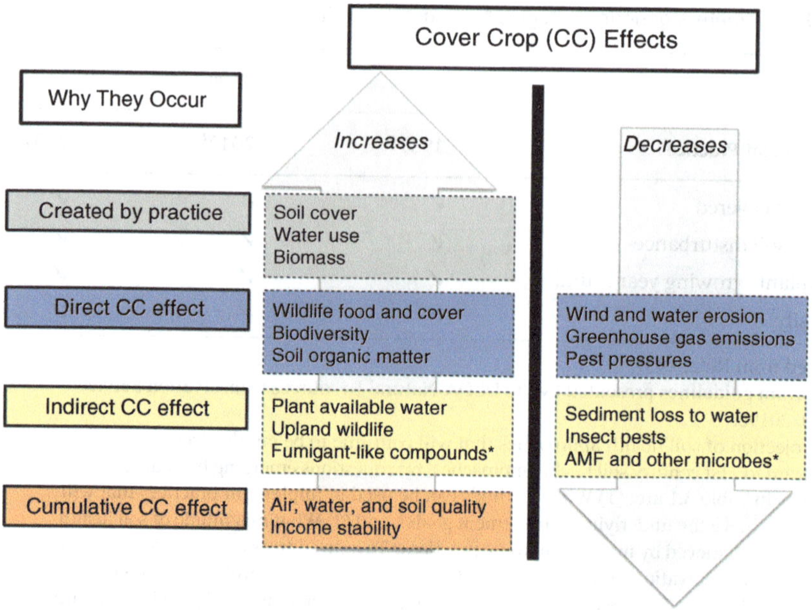


**Fig 3.2** Utilizing cover crops as a conservation practice to improve soil health can have many different effects and interactions. *Source:* USDA-NRCS (https://www.nrcs.usda.gov/wps/PA_NRCSConsumption/download?cid = stelprdb1270377&ext=pdf).

acreage distributions, harvest practices, storage requirement and ultimately market availability and potential economic returns, especially if new or non-traditional crop rotations are introduced. Such changes are not trivial and can involve substantial time, effort, and financial costs, and may not be feasible or practical due to a range of climatic, social, or financial factors (Carter, 2019; Findlater et al., 2019; Giller et al., 2009; Janzen, 2001).

Recognizing these and other implementation challenges, it then becomes very important to consider the second question– how much of a soil health benefit can be achieved if those practices are implemented? The key phrase in that question is "how much!" Once again, there is no simple answer because every situation is site-specific with regard to comparison groups, scale of production or implementation, cost and sensitivity of analytical methods, and the degree to which a biological, chemical, physical or overall soil health change is measurable. The latter must also consider whether the change is of statistical (*i.e.*, *p*-value) or practical value. We will now focus briefly on each of those concepts to explore how different decisions and actions will ultimately determine cost to benefit ratios for each of the potential soil health approaches.

## Comparison groups

When documenting the specific benefit of a management practice, the potential magnitude of this change will depend on both inherent and dynamic soil properties (Fig. 3.3). For example, when considering tillage, visualizing a disturbance continuum can be helpful because humans have the ability to shape soils from a range tools and techniques, including tillage (Reicosky, 2015) and drainage (Dinnes et al., 2002; Skaggs et al., 1994). Similarly, when comparing a range of management practices, selection of two practices with major differences (*e.g.*, moldboard plow *vs* native prairie grass) will likely show more significant differences due to contrasting levels of soil disruption, plant species, and external inputs (Veum et al., 2014; Veum et al., 2015). Furthermore, the way in which the comparisons are statistically conducted can definitely influence "how much" of an effect was present and detectable (van Es and Karlen, 2019; Roper et al., 2017).

## Scale

The scale of measurement is an important consideration for soil health assessments because soil resources can have very different properties when viewed across an individual field, across an entire landscape, as a core representing the soil profile, or as a small sample prepared for one or more analytical measurements. Within-field variability is another factor that can make soil health assessment useful or futile. For example, changes in SOM could reflect either an increasing level or an unintended over-sampling from soil series within the field

**Figure 3.3** Soil health documentation must recognize inherent (left) and dynamic (right) soil properties. (Photo Credit: Gary Radke, USDA ARS).

that tend to have higher SOM levels than others. Slope is another common factor in agricultural fields that helps explain differences in SOM (Ladoni et al., 2016; Ontl et al., 2015). Also, if samples are collected from within a row, are the measurements applicable to the entire field or only the 25% of the field that is within a crop row? These questions are relevant not only for SOM but all potential soil health indicators. For example, in a study of carbon dioxide ($CO_2$) flux from no-till fields, Kaspar and Parkin (2011) estimated each field consisted of 25% rows, 45% untracked inter-rows, and 30% tracked inter-rows, thus highlighting the importance of controlling traffic patterns to ensure that only a small portion of each field is disturbed. Quantifying such fine scale variations in soil properties is an important component of soil health assessment since the results may help identify new crop management practices that can improve all functional zones in the soil (Williams et al., 2016).

## Analytical Methods

After selecting the soil function(s) for which indicator comparisons are of interest and the appropriate scale for making the comparisons, the third factor to consider is which method should be used to measure the important or critical changes? These three factors (topic, scale, and methods) are core scientific questions within any field of study. This book and numerous others document that a range of soil sampling and analytical methods exist (Dane and Topp, 2002; Dick, 2011; Sparks et al., 1996, Ulery and Drees, 2008). Therefore, the term "soil health test" can refer to a multitude of in-field, laboratory, or even remote sensing techniques for quantifying or documenting a specific soil function or indicator of that function (Table 3.3). This then fuels ongoing discussions with regard to the utility or futility of soil health assessment and which soil physical, chemical, and/or biological properties should be documented (Derner et al., 2018; Doran et al., 1994; Doran and Jones, 1996; Elliott et al., 1997; Schindelbeck et al., 2008; Stone et al., 2016).

Common soil health tests (Table 3.3) include in-field assessments with scorecards or portable soil test kits that can be used to evaluate soils visually and interactively. A more involved type of in-field assessment can be achieved through the installation of various analytical instruments including edge-of-field or end-of-drainage-tile samplers for quantifying soil health impacts on water quality. Assessments can also be made by sending soil samples to a commercial testing laboratory, or by participating in research projects such as the National Corn Grower sponsored Soil Health Partnership, the Soil Health Institute's national soil health evaluation, or one of many state or national NRCS soil health programs. A third category of soil health tests emerging through technological advances is the

**Table 3.3** Categories of soil health tests, each with unique characteristics but a common goal.

| Type of test | Characteristics | Common Goal |
|---|---|---|
| On-farm or in-field | Portable, generally quite simple, qualitative, interactive, provide general contrasts | Successfully identify if soil properties change |
| Commercial laboratory | Rapid and high throughput, primarily focused on chemical indicators, with a few physical and biological measurements, generally group responses in categories | |
| Research projects | More precise but often very slow turn-around, capable of identifying fine-scale differences, difficult to generalize, specific methods may vary | |

use of remote sensing which has the benefit of enabling more frequent assessments of several soil functions, but the challenge of amassing high volumes of data which require more sophisticated storage and interpretation algorithms (Mulder et al., 2011; Shoshany et al., 2013).

The three categories of soil health tests listed in Table 3.3 thus serve different purposes. On-farm or in-field qualitative tests are generally used to build an awareness of what soil is, how it forms, and the types of functions (sustaining productivity, filtering and buffering, controlling water entry, retention and release). Commercial or research-based laboratory tests generally require collection of numerous samples and sending them to a separate location for analysis. Also, since the farmer/landowner/interested individual is often not the one collecting or analyzing the samples, there can be disconnects or even lack of communication between the person providing analytical data and the one who will ultimately use the information to modify decisions and/or change soil and crop management practices. Regardless of the specific type of test, a very important cultural change associated with development of soil health concepts has been the act of bringing people together, often in the field, to evaluate the soil and thus better understand benefits that often cannot be easily seen through printouts of laboratory data. For example, an area prone to erosion can often be documented more easily by evaluating the slope, amount of groundcover, and presence of ephemeral or permanent gullies than looking at data showing soil texture, SOM, or fertility changes.

Another difference among the three categories of soil health tests (Table 3.3) is that they are generally applied at different scales. For laboratory tests extensive pre-processing, such as sieving, grinding or sub-sampling, will often be required as samples

**Figure 3.4** Potential scales at which soil health indicators can be assessed. (Photo credit: Gary Radke, USDA ARS).

are prepared for analysis using various analytical instruments. In contrast, on-farm or in-field tests can often help producers recognize impacts of past and current soil management decisions within a field or landscape more easily than viewing multiple pages of laboratory data. Furthermore, if the small amount of soil submitted for laboratory analysis is not accompanied by an appropriate amount of metadata (*i.e.*, data about the samples, site, and analytical methods) it may be impossible to fully capture what can be seen within-field by the naked eye. Another visual assessment technique that has become more common since the emergence of soil health conferences and field days is the increased use and familiarity of soil pits to show producers how their crop production practices are influencing plant root systems and soil structure. Rainfall simulators are also used in soil health field days to demonstrate the benefits of keeping soil covered and slake tests are used to show the importance of stable soil aggregates and SOM as compared to cloddy, compacted soils.

Two important realities of soil health assessment, regardless of the specific test being used, are the recognition and improved understanding of soil function. Having that knowledge enables NRCS and other consultants to help producers evaluate several soil-related natural resource concerns (Table 3.4) and to recognize that although any of the three types of soil health tests could be used, some may be more useful for determining if a specific concern is present and

**Table 3.4** Potential soil health tests for evaluating various natural resource concerns.

| | **Most common type of soil health assessment** | | |
|---|---|---|---|
| **Natural resource concern** | **In-field** | **Commercial lab** | **Research lab** |
| Erosion | ✓ | | |
| SOM depletion | ✓ | ✓ | ✓ |
| Elevated salts | ✓ | ✓ | ✓ |
| Excess water | ✓ | | |
| Insufficient water | ✓ | | |
| Pesticide transport | | ✓ | ✓ |
| Excess pathogens | ✓ | ✓ | ✓ |
| Heavy metals | | ✓ | ✓ |
| Sediment in surface waters | ✓ | ✓ | ✓ |
| Elevated water temperatures | ✓ | | |
| Degraded plant condition | ✓ | | |
| Wildfire hazard | ✓ | | |
| Odors | ✓ | | |

subsequently how to address it. Therefore, it is important to recognize that the assignments we list in Table 3.4 simply represent common measurements and that other tests could easily be used.

It is also important to recognize that each type of soil health test will have variability associated with the end result. Furthermore, documenting changes in soil properties is also challenging by the fact that soils are inherently variable. A test with high variability may complicate its use to quantify changes due to management simply because any true changes become lost within the larger variability of the test itself. Selecting the appropriate type of test and scale at which to use it is therefore an ongoing question in soil science. Thus as land management practices change, soil health measurements may also have to change to detect subtle changes in soil properties or functions.

The living, dynamic nature of soil resources contributes to what some consider the futility of soil health assessment. For example, depending on site-specific field properties, the number of soil samples required for a meaningful soil health measurement can vary widely (Cambardella et al., 1994; Hurisso et al., 2018; Ladoni et al., 2015; Morrow et al., 2016; Necpálová et al., 2014). Also, although there are several types of qualitative and quantitative tests that can be used to analyze soil health properties, the decision on which approach to use will ultimately depend on the type of questions that are to be addressed. For example, sediment in surface waters can be detected using in-field techniques by simply noting the presence of soil particles in water being collected as run-off from a specific area. These in-field tests could be made more quantitative by documenting the amount of sediment per unit volume of water if a known volume is collected, the water is evaporated, and the remaining amount of sediment is weighed. However, if the goal is to determine the concentration of a specific element or chemical in the run-off water, the surface water samples that are being collected will have to be sent to a commercial or research laboratory where analytical tests beyond the scope of an in-field test can be made. Or, substantial in-field or edge-of-field instrumentation will need to be installed to quantify these concentrations.

## Degrees of Change

Opinions regarding the utility or futility of soil health assessment are often based on the challenges of documenting benefits from soil health approaches which are highly dependent on the question of interest, type of test used, and scale at which the test is applied. We advocate that to be considered soil health research, assessments must include soil physical, chemical, and biological properties, although some would argue that focusing on one particular soil property is sufficient. We consider that latter approach to be soil physical, soil chemical, or soil biological health and not soil health per se.

Other chapters in this book provide a more thorough discussion and specific information regarding individual soil health assessment methods. The examples included herein are simply intended to be illustrative of how soil health benefits have been documented in the scientific literature. Our goal is to highlight the types of soil system comparisons that have been made, including those examining different types of cropping systems, large scale changes associated with a disturbance continuum, and subtle changes in soil properties over time. Obviously, these are not the only types of comparisons within the soil health literature, but they were selected to illustrate the types of challenges associated with documenting soil health benefits.

Soils managed using no-tillage with the addition of cover crops are expected to have better soil health properties than tilled soils without cover crops because the former results in greater SOM, higher soil enzyme activities, and more stable soil aggregation. Furthermore, soils managed under no-till with cover crops fulfill many of the goals from Table 3.2, including maintaining soil cover, reducing soil disturbance, extending the time when vegetation is growing for as long as possible, and diversifying plant species across the landscape. However, these potential soil health benefits are not guaranteed. If cover crop establishment is poor for several consecutive years because of weather patterns (*e.g.*, unusually early freezing or substantial drought conditions), the magnitude of soil change could be quite limited. This is precisely the situation experienced by VeVerka et al. (2019) in a study of Missouri claypan soils across four watersheds for 3 yr. They were able to detect expected depth differences in soil properties, with surface samples having greater soil biological activity, as indicated by higher enzyme and organic matter values. However, this trend did not occur for glucosaminidase, since quantities of that enzyme actually increased with depth. Overall, the expected soil health indicator improvement expected for no-till plus cover crops in comparison to tilled, no cover crop treatments did not occur presumably because of unfavorable growing conditions.

Another common short-term soil health evaluation is to compare two widely varying production systems from opposite ends of a disturbance continuum (*e.g.*, perennial grassland *vs.* a tilled field). Transitioning from grassland to a tilled field typically causes shifts in soil biological activities, sometimes within the first month after tillage commences. In Texas, Cotton and Acosta-Martínez (2018) documented a 52% decrease (505 to 241 mg $kg^{-1}$ soil) in soil microbial biomass in the top 10 cm of the soil profile. After the first growing season, soil organic carbon (SOC) within the top 30 cm of the profile declined by an average of 20%, although the decline in the surface 10 cm (11.60 *vs.* 7.28 g SOC $kg^{-1}$ soil) was substantially greater than within the 10- to 30-cm increment (6.76 *vs.* 6.17 g SOC $kg^{-1}$ soil). This was not unexpected since in most soil health studies, soil property changes are greater near the soil surface because this portion of the profile is most directly affected by changes in tillage and plant root conditions.

Soil management practices that increase microbial food supplies and reduce disturbance to their habitats will tend to have greater soil microbial activity than soils with limited food sources (*i.e.*, SOC), due to low crop residue or root carbon inputs, excessive crop residue removal, or excessive tillage. Sustaining adequate soil microbial activity is important because biologically mediated processes such as nutrient cycling and SOM dynamics are critical components of several soil ecological functions (Dick, 1992). For example, in comparisons between conservation reserve program (CRP) fields and active croplands, Li et al. (2018) found that fungal abundance increased in proportion to the length of time since CRP practices were implemented. They found increases in fungal abundance up to 15 yr after establishment, followed by decreases relative to bacterial abundance. The shifts in microbial community composition were attributed to historical soil conditions, abiotic factors and climate properties. Soils in CRP had less stressed soil microbial communities as indicated by fatty-acid methyl ester biomarker (FAME) profiles, in which the ratio of saturated to mono-unsaturated fatty-acids decreased (Li et al., 2018). The mono-unsaturated fatty-acids indicated active metabolic processes were occurring in the soil compared to higher saturated fatty-acid profiles that are indicative of slower metabolic processes which can occur when water or nutrients are in short supply for microbes.

## Soil Health Limitations

There are at least three related components that can limit the utility of soil health research and implementation efforts. First are the logistical limitations including the cost of a project, access to samples or instrumentation, and time for the study to be conducted. Next, are the philosophical limitations that can occur when an assessment project is designed, especially regarding the scope and questions of interest. These types of limitations generally occur if someone considers the project's approach to be insufficient to answer what is often potentially a broader or different question. Logistical and philosophical limitations do overlap. For example, consider the two questions: "Can we measure what we want to measure"? and "Why do we want to conduct the measurement"? A third important limitation involves the number of management changes that are needed to fully execute a comprehensive soil health management approach.

Phosphorus loss from agricultural fields provides an excellent example of this third limitation. Growing cover crops and reducing tillage are often promoted as soil health improving practices because of their impact on SOC, but they may not be sufficient to reduce P losses from agricultural fields. Since either high soil-test P levels, excessive P applications through fertilizer or animal manure, or high soluble P in senescing cover crop vegetation can all contribute to increased soluble P runoff, soil health management practices focused on this problem must be coupled with changes in the way P is applied to the fields. This could easily involve new materials, timing of application, and/or equipment (Duncan et al., 2019). Therefore, an effective soil health research project may require substantial changes to all dynamic soil properties and associated management practices, which from a practical perspective can be a limitation when multiple fields and/or producers are involved.

Other potential limitations to useful soil health research and technology transfer include factors such as producer interest, economic limitations, time requirements, and the magnitude of change needed with regard to soil and crop management practices and/or desired with regard to soil properties. The utility, however, is emphasized by the numerous potential endpoints that exist, especially when balancing productivity with a broader environmental perspective. Is the ultimate endpoint of improved soil health increased yield, long-term sustainability, water quality, economic viability, community development, or all of these goals? The length of time for which soil health indicators must be tracked and whether or not changes can be documented will be determined by the ultimate goal(s). This also determines the magnitude and type of change that must be measured. Without any doubt, research studies can document findings that are both statistically significant and practically important. However, depending on (1) how changes are measured, such as with an in-field test, commercial, or research laboratory test, (2) the inherent soil variability and (3) the analytical soil test variability, one or more of those factors can potentially mask any true soil health effects. It is not surprising, therefore, that all of these challenges (*i.e.*, endpoints, time, magnitude of change) reflect various trade-offs. Research projects tend to be funded for relatively short periods of time, often measured in two to five year increments, research budgets are not unlimited, and every sample that needs to be analyzed requires careful collection, appropriate preparation and adequate processing time. Obviously, these challenges are not unique to soil health research, but recognizing them may help diffuse some of the discussion between those who view the efforts as either useful or futile.

## Conclusions

Documenting benefits from soil health approaches first requires defining what is the benefit of interest and then selecting ways to measure and document the response. The principles associated with soil health are not new as evident by centuries of soil management, conservation, condition, tilth, quality, and other terms. Among the well-known and generally accepted approaches for improving soil health are the goals of keeping the soil covered, reducing disturbance, maintaining plants year-round, and diversifying the mix of plant species. Implementing these goals can increase the quantity of plant residues and root exudates returned to the soil, boost microbial activity, and ultimately lead to a cascade of soil improvements, including increased SOM, more stable soil aggregation, and efficient nutrient cycling. How well these benefits can be documented depends on the magnitude of change (generally determined by inherent soil properties and/or initial conditions) as well as the type of soil health test selected (*i.e.*, in-field, commercial or research laboratory, remote sensing), and the scale at which comparisons are to be made and meaningful (*i.e.*, from landscapes down to finely sieved and crushed soil samples). Interactions between inherent and dynamic soil properties can also make documenting soil health benefits difficult, since spatial and temporal variability can mask potential changes associated with new soil and crop management practices, such as annual cover crop establishment, which must be given adequate time for measurable effects to occur. Extreme weather conditions, such as too much or too little rainfall, early or late frost, or above normal temperatures can hinder the effectiveness of new or alternative management systems and prevent them from becoming established and changing soil properties in subsequent years. Without question, researchers have documented numerous benefits from soil health approaches. As the concept evolves, the core questions of defining what constitutes an important benefit and selecting reproducible methods to measure that benefit will remain a constant challenge and an important research goal.

## References


Andrews, S.S., Karlen, D.L., and Cambardella, C.A. (2004). The soil management assessment framework: A quantitative soil quality evaluation method. *Soil Sci. Soc. Am. J.* 68(6), 1945–1962. doi:10.2136/sssaj2004.1945

Bennett, H.H., and Chapline, W.R. (1928). *Soil erosion: A national menace.* United States Department of Agriculture Circular 33. Washington, D.C.: United States Government Printing Office.

Cambardella, C.A., Moorman, T.B., Novak, J.M., Parkin, T.B., Karlen, D.L., Turco, R.F., and Konopka, A.E. (1994). Field-scale variability of soil properties in central Iowa soils. *Soil Sci. Soc. Am. J.* 58(5), 1501–1511. doi:10.2136/sssaj1994.03615995005800050033x

Carter, A. (2019). "We don't equal even just one man": Gender and social control in conservation adoption. *Soc. Nat. Resour.* 32(8), 893–910. doi:10.1080/0894192 0.2019.1584657

Cotton, J., and Acosta-Martínez, V. (2018). Intensive tillage converting grassland to cropland I mmediately reduces soil microbial community size and organic carbon. *Agricultural and Environmental Letters* 3:180047. doi:10.2134/ael2018.09.0047

Dane, J.H., and Topp, C.G., (eds.). (2002). Methods of soil analysis: Part 4 physical methods. SSSA Book Ser. 5.4. Madison, WI: SSSA. doi:10.2136/sssabookser5.4.

Derner, J.D., Smart, A.J., Toombs, T.P., Larsen, D., McCulley, R.L., Goodwin, J., Sims, S., and Roche, L.M. (2018). Soil health as a transformational change agent for us grazing lands management. *Rangeland Ecology & Management* 71(4), 403–408. doi:10.1016/j.rama.2018.03.007

Diamond, J. (2011). Collapse: How Societies Choose to Fail or Succeed. Revised ed. *Penguin Books.*

Dick, R.P. 1992. A review: Long-term effects of agricultural systems on soil biochemical and microbial parameters. *Agric. Ecosyst. Environ.* 40:25–36. doi:10.1016/0167-8809(92)90081-L

Dick, R.P., (ed.). (2011). *Methods of soil enzymology*. SSSA Book Ser. 9. Madison, WI: SSSA. doi:10.2136/sssabookser9

Dinnes, D.L., Karlen, D.L., Jaynes, D.B., Kaspar, T.C., Hatfield, J.L., Colvin, T.S., and Cambardella, C.A. (2002). Nitrogen management strategies to reduce nitrate leaching in tile- drained midwestern soils. *Agronomy Journal* 94(1), 153–171. doi:10.2134/agronj2002.0153

Doran, J.W., Coleman, D.C., Bezdicek, D.F., Stewart, B.A. (eds.). (1994). *Defining soil quality for a sustainable environment*. SSSA Spec. Publ. 35. Madison, WI: SSSA and ASA.

Doran, J.W., and Jones, A.J., (eds.). (1996). *Methods for assessing soil quality*. SSSA Spec. Publ. 49. Madison, WI: SSSA.

Duncan, E.W., D.L. Osmond, A.L. Shober, L. Starr, P. Tomlinson, J.L. Kovar, T.B. Moorman, H.M. Peterson, N.M. Fiorellino, and K. Reid. 2019. Phosphorus and soil health management practices. *Agricultural and Environmental Letters* 4:1900014. doi:10.2134/ael2019.04.0014

Elliott, E.T., Pankhurst, B.E., Doube, C.E., and Gupta, V.V.S.R. (1997). Rationale for developing bioindicators of soil health. In: C. Pankhurst, (eds.), *Biological indicators of soil health* (p. 49–78). Wallingford, U.K.: CSIRO Division of Soils. CABI Publishing.

Findlater, K.M., Satterfield, T., and Kandlikar, M. (2019). Farmers' risk-based decision making under pervasive uncertainty: Cognitive thresholds and hazy hedging. *Risk Anal.* 39(8), 1755–1770. doi:10.1111/risa.13290

Fream, W. (1890). Tilth. p. 95–100. In W. Fream, *Soils and their properties*. London: George Bell & Sons.

Gebhart, D.L., Johnson, H.B., Mayeux, H.S., and Polley, H.W. (1994). The CRP increases soil organic-carbon. *J. Soil Water Conserv.* 49(5), 488–492.

Giller, K.E., Witter, E., Corbeels, M., and Tittonell, P. (2009). Conservation agriculture and smallholder farming in Africa: The heretics' view. *Field Crops Res.* 114(1), 23–34. doi:10.1016/j.fcr.2009.06.017

Heller, M.C. and Keoleian, G.A. (2003). Assessing the sustainability of the US food system: A life cycle perspective. *Agricultural Systems* 76(3), 1007–1041. doi:10.1016/S0308-521X(02)00027-6521X(02)00027-6

Helms, D. (1991). Two centuries of soil conservation. *OAH Magazine of History* 5(3), 24–28. doi:10.1093/maghis/5.3.24

Hillel, D. (1991). *Out of the earth: Civilization and the life of the soil.* Oakland, CA: University of California Press

Hobbs, P.R. (2007). Conservation agriculture: What is it and why is it important for future sustainable food production? *J. Agric. Sci.* 145, 127–137. doi:10.1017/S0021859607006892

Hurisso, T.T., Culman, S.W., and Zhao, K. (2018). Repeatability and spatiotemporal variability of emerging soil health indicators relative to routine soil nutrient tests. *Soil Sci. Soc. Am. J.* 82, 939–948. doi:10.2136/sssaj2018.03.0098

Janzen, H.H. (2001). Soil science on the Canadian prairies– Peering into the future from a century ago. *Can. J. Soil Sci.* 81, 489–503. doi:10.4141/S00-054

Janvier, C., Villeneuve, F., Alabouvette, C., Edel-Hermann, V., Mateille, T., and Steinberg, C. (2007). Soil health through soil disease suppression: Which strategy from descriptors to indicators? *Soil Biol. Biochem.* 39(1), 1–23. doi:10.1016/j.soilbio.2006.07.001

Karlen, D.L., Mausbach, M.J., Doran, J.W., Cline, R.G., Harris, R.F., and Schuman, G.E. (1997). Soil quality: A concept, definition, and framework for evaluation. *Soil Sci. Soc. Am. J.* 61(1), 4–10. doi:10.2136/sssaj1997.03615995006100010001x

Karlen, D.L., Veum, K.S., Sudduth, K.A., Obrycki, J.F., and Nunes, M.R. (2019). Soil health assessment: Past accomplishments, current activities, and future opportunities. *Soil Tillage Res.* 195, 104365. doi:10.1016/j.still.2019.104365

Kaspar, T.C., and Parkin, T.B. (2011). Soil carbon dioxide flux in response to wheel traffic in a no-till system. *Soil Sci. Soc. Am. J.* 75(6), 2296–2304. doi:10.2136/sssaj2011.0129

Keen, B.A. (1931). *The physical properties of the soil. Rothamsted monograph on Agricultural Science.* Ser. London: Longmans, Green and Co.

King, F.H. 1911. *Farmers of forty centuries.* Madison, Wisconsin: F.H. King.

Ladoni, M., Basir, A., and Kravchenko, A. (2015). Which soil carbon fraction is the best for assessing management differences? A statistical power perspective. *Soil Sci. Soc. Am. J.* 79(3), 848–857. doi:10.2136/sssaj2014.10.0426

Ladoni, M., Basir, A., Robertson, P.G., and Kravchenko, A.N. (2016). Scaling-up: Cover crops differentially influence soil carbon in agricultural fields with diverse topography. *Agric. Ecosyst. Environ.* 225, 93–103. doi:10.1016/j.agee.2016.03.021

Li, C., Fultz, L.M., Moore-Kucera, J., Acosta-Martínez, V., Horita, J., Strauss, R., Zak, J., Calderon, F., and Weindorf, D. 2017. Soil carbon sequestration potential in semi-arid grasslands in the Conservation Reserve Program. *Geoderma* 294, 80–90. doi:10.1016/j.geoderma.2017.01.032

Li, C., Fultz, L.M., Moore-Kucera, J., Acosta-Martínez, V., Kakarla, M., and Weindorf, D.C. (2018). Soil microbial community restoration in Conservation Reserve Program semi-arid grasslands. *Soil Biol. Biochem.* 118, 166–177. doi:10.1016/j.soilbio.2017.12.001

Lowdermilk, W.C. (1953). *Conquest of the land through 7000 years*. Washington, D.C.: USDA Soil Conservation Service, U.S. Department of Agriculture.

Magdoff, F., and van Es, H. (2009). *Building soils for better crops: Sustainable soil management*, 3rd ed. Sustainable Agriculture Research and Education, Handbook Series Book 10. College Park, MD: Sustainable Agriculture Research and Education.

McBratney, A., Field, D.J., and Koch, A. (2014). The dimensions of soil security. *Geoderma* 213, 203–213. doi:10.1016/j.geoderma.2013.08.013

Mena Mesa, N., Ruiz-Vega, J., Funes-Monzote, F.R., Carrillo-Rodriguez, J.C., and Velasco-Velasco, V. (2014). Indicators of agroecological sustainability of three tillage systems for maize (*Zea mays* L.) production. *Agroecology and Sustainable Food Systems* 38(4), 410–426. doi:10.1080/21683565.2013.870626

Moebius-Clune, B.N., Moebius-Clune, D.J., Gugino, B.K., Idowu, O.J., Schindelbeck, R.R., Ristow, A.J., van Es, H.M., Thies, J.E., Shayler, H.A., McBride, M.B., Kurtz, K.S.M., Wolfe, D.W., and Abawi, G.S. (2016). *Comprehensive assessment of soil health–The Cornell framework*, edition 3.2, Geneva, NY: Cornell University.

Montgomery, D.R. (2007). *Dirt: The erosion of civilizations*, 2nd ed. Berkeley, CA: University of California Press.

Morrow, J.G., Huggins, D.R., Carpenter-Boggs, L.A., and Reganold, J.P. (2016). Evaluating measures to assess soil health in long-term agroecosystem trials. *Soil Sci. Soc. Am. J.* 80, 450–462. doi:10.2136/sssaj2015.08.0308

Mulder, V.L., de Bruin, S., Schaepman, M. (2011). The Use of Remote Sensing in Soil and Terranin Mapping- A Review. *Geoderma* 162 (1–2): 1–19, https://doi.org/10.1016/j.geoderma.2010.12.018

Necpálová, M., Anex, Jr., R.P., Kravchenko, A.N., Abendroth, L.J., Del Grosso, S.J., Dick, W.A., Helmers, M.J., Herzmann, D., Lauer, J.G., Nafziger, E.D., Sawyer, J.E., Scharf, P.C., Strock, J.S., and Villamil, M.B. (2014). What does it take to detect a change in soil carbon stock? A regional comparison of minimum detectable difference and experiment duration in the north central United States. *J. Soil Water Conserv.* 69(6), 517–531. doi:10.2489/jswc.69.6.517

Ontl, T.A., Cambardella, C.A., Schulte, L.A., and Kolka, R.K. (2015). Factors influencing soil aggregation and particulate organic matter responses to bioenergy crops across a topographic gradient. *Geoderma* 255–256, 1–11. doi:10.1016/j.geoderma.2015.04.016

Rasul, G., and Thapa, G.B. (2004). Sustainability of ecological and conventional agricultural systems in Bangladesh: An assessment based on environmental, economic and social perspectives. *Agric. Syst.* 79(3), 327–351. doi:10.1016/S0308-521X(03)00090-8

Reicosky, D.C. (2015). Conservation tillage is not conservation agriculture. *J. Soil Water Conserv.* 70(5), 103–108. doi:10.2489/jswc.70.5.103A

Roper, W.R., Osmond, D.L., Heitman, J.L., Wagger, M.G., and Reberg-Horton, S.C. (2017). Soil health indicators do not differentiate among agronomic management systems in North Carolina soils. *Soil Sci. Soc. Am. J.* 81(4), 828–843. doi:10.2136/sssaj2016.12.0400

Rule, G.K. (1937).*Conserving corn belt soils.* Farmers' Bulletin No. 1795. Washington, D.C.: U.S. Gov. Print. Office.

Schindelbeck, R.R., van Es, H.M., Abawi, G.S., Wolfe, D.W., Whitlow, T.L., Gugino, B.K., Idowu, O.J., and Moebius-Clune, B.N. (2008). Comprehensive assessment of soil quality for landscape and urban management. *Landsc. Urban Plan.* 88(2-4), 73–80. doi:10.1016/j.landurbplan.2008.08.006

Schnepf, M., and Cox, C. (2006). *Environmental benefits of conservation on cropland: The status of our knowledge.* Ankeny, IA: Soil and Water Conservation Society.

Shoshany, M., Goldshleger, N., Argaman, E., Chudnovsky, A. (2013). Monitoring of Agricultural Soil Degradation by Remote-Sensing Methods: A Review. *International Journal of Remote Sensing* 34 (17): 6152–6181, https://doi.org/10.1080/01431161.2013.793872

Skaggs, R.W., Breve, M.A., and Gilliam, J.W. (1994). Hydrologic and water-quality impacts of agricultural drainage. *Crit. Rev. Environ. Sci. Technol.* 24(1), 1–32. doi:10.1080/10643389409388459

Sparks, D.L., A.L. Page, P.A. Helmke, and R.H. Loeppert, editors. 1996. *Methods of soil analysis part 3—chemical methods.* SSSA Book Ser. 5.3. Madison, WI: SSSA, ASA. doi:10.2136/sssabookser5.3.

Steffan, J.J., Brevik, E.C., Burgess, L.C., and Cerdà, A. (2017). The effect of soil on human health: A review. *Eur. J. Soil Sci.* 69(1), 159–171. doi:10.1111/ejss.12451

Stoll, S. (2003). *Larding the lean earth: Soil and society in nineteenth-century America.* New York: Hill and Wang.

Stone, D., Ritz, K., Griffiths, B.G., Orgiazzi, A., and Creamer, R.E. (2016). Selection of biological indicators appropriate for European soil monitoring. *Appl. Soil Ecol.* 97, 12–22. doi:10.1016/j.apsoil.2015.08.005

Ulery, A.L., and Drees, R., (eds.). (2008). *Methods of soil analysis: Part 5 mineralogical methods.* SSSA Book Ser. 5.5. Madison, WI: SSSA. doi:10.2136/sssabookser5.5

USDA NRCS. (2019a). *National conservation practice standards.* Washington, D.C.: USDA-NRCS. https://www.nrcs.usda.gov/wps/portal/nrcs/main/national/technical/cp/ncps/

USDA NRCS. 2019b. *Soil health management.* Washington, D.C.: USDA-NRCS. https://www.nrcs.usda.gov/wps/portal/nrcs/main/soils/health/mgnt/

van Es, H.M., and Karlen, D.L. (2019). Reanalysis validates soil health indicator sensitivity and correlation with long-term crop yields. *Soil Sci. Soc. Am. J.* 83(3), 721–732. doi:10.2136/sssaj2018.09.0338

Veum, K.S., K.W. Goyne, R.J. Kremer, Miles, R.J., and Sudduth, K.A. (2014). Biological indicators of soil quality and soil organic matter characteristics in an agricultural management continuum. *Biogeochemistry* 117(1), 81–99. doi:10.1007/s10533-013-9868-7

Veum, K.S., Kremer, R.J., Sudduth, K.A., Kitchen, N.R., Lerch, R.N., Baffaut, C., Stott, D.E., Karlen, D.L., and Sadler, E.J. (2015). Conservation effects on soil quality indicators in the Missouri Salt River Basin. *J. Soil Water Conserv.* 70(4), 232–246. doi:10.2489/jswc.70.4.232

VeVerka, J.S., Udawatta, R.P., and Kremer, R.J. (2019). Soil health indicator responses on Missouri claypan soils affected by landscape position, depth, and management practices. *J. Soil Water Conserv.* 74(2), 126–137. doi:10.2489/jswc.74.2.126

Williams, A., Kane, D.A., Ewing, P.M., Atwood, L.W., Jilling, A., Li, M., Lou, Y., Davis, A.S., Grandy, A.S., Huedi, S.C., Hunter, M.C., Koide, R.T., Mortensen, D.A., Smith, R.G., Snapp, S.S., Spokas, K.A., Yannarell, A.C., and Jordan, N.R. (2016). Soil functional zone management: A vehicle for enhancing production and soil ecosystem services in row-crop agroecosystems. *Front. Plant Sci.* 7, 1–65. doi:10.3389/fpls.2016.00065

Wolde, B. Lal, P., Alavaapati, J., Burli, P., and Iranah, P. (2016). Soil and water conservation using the socioeconomics, sustainability concerns, and policy preference for residual biomass harvest. *J. Soil Water Conserv.* 71(6):476–483. doi:10.2489/jswc.71.6.476

# 4

# Metadata: An Essential Component for Interpreting Soil Health Measurements

*Jane M.-F. Johnson and Maysoon M. Mikha*

Soil health, also called soil quality, has both inherent edaphic components and dynamic properties, which interact with management and climatic affects (Karlen, Ditzler, & Andrews, 2003). Inherent soil characterization has been an integral part of potential land use assessments for decades. In contrast, the concept of soil health has evolved with a primary focus on agronomic productivity but including effects on plant nutrition and ultimately the effects on human health (Robinson et al., 2017).

Discussions of soil assessment often focus on the suitability of indicators or attribute data but fail to adequately address the metadata (Bünemann et al., 2018). Attribute data reflects the anticipated or actual response to management practices and are related to indices of soil health (Askari & Holden, 2015). However, metadata provides the information needed to establish a context for accurately interpreting indicator data, thereby facilitating synthesis and integration of experimental results across time and/or space.

Soil health or perhaps even more importantly changes in soil properties (i.e., measured attribute data) that describe soil health inherently require a geospatial and temporal context. Spatial components include the site description, topographical data, and inherent soil parameters. Temporal data describe when

Abbreviations: DET, data entry template; GRACEnet, Greenhouse gas Reduction through Agricultural Carbon Enhancement network; SOC, soil organic carbon.

*Soil Health Series: Volume 1 Approaches to Soil Health Analysis*, First Edition.
Edited by Douglas L. Karlen, Diane E. Stott, and Maysoon M. Mikha.

the data were collected. This is important because granularity of temporal measurements differ at vastly different scales ranging from nearly instantaneous to seasonal or even yearly depending on the parameter being evaluated. Metadata should also include a wide range of edaphic parameters.

Furthermore, an excellent example of both the importance and use of metadata can be found by examining the USDA–ARS Greenhouse Reduction through Agricultural Enhancement network (GRACEnet) protocols. GRACEnet was developed to answer questions related to agricultural soil C stocks in the context of greenhouse gas mitigation (Del Grosso et al., 2013; Jawson, Shafer, Franzluebbers, Parkin, & Follett, 2005; Liebig, Franzluebbers, & Follett, 2012, p. 547). When organizing this large, complex, multisite project, empirical data collectors, data users, data synthesizer, and modelers worked together to identify the metadata they considered necessary to interpret and model the attribute data. The output of GRACEnet organization meetings established a set of metadata and indicator (measurement) data that can be used to assess the impact of management on soil organic C (SOC). The linkage between SOC and soil health through metadata was first emphasized by Doran (2002), because it is the metadata that can help to logically interpret SOC changes and their effect on soil health.

The GRACEnet effort developed a data entry template (DET) to assist in collating meta and attribute data from multiple users (Del Grosso et al., 2013). The DET has been modified to meet a range of experimental purposes (Delgado et al., 2016; Del Grosso et al., 2013; Liebig et al., 2016) but recommends the characterization or metadata collected that has to be conserved: site description, soil characterization, management, climate, and sampling and analysis methods (Table 4.1). The DET thus provides a prime example of the metadata needed to adequately interpret soil parameters when striving to describe or assess soil health.

## Methods and Frequency

### Site Description

Site description includes high-level information needed to understand whether soil properties and processes are static or change slowly (Liebig, Varvel, & Honeycutt, 2010) at a given site. This includes location, soil characteristics, experimental objectives and design, and climatic and weather data.

Location may be described using latitude, longitude, and elevation and may include a city, state, and country reference and topographic descriptors. Other site descriptive data that can be useful are the Major Land use Resource Area

**Table 4.1** Metadata related to temporal and spatial properties of soil health.

| Site description | | | | |
|---|---|---|---|---|
| **Location** | **Soil characterization** | **Experimental descriptors** | **Climatic data** | **Weather** |
| City, state, country | soil taxonomic description | experimental design | mean annual temperature | nearest weather station location |
| Latitude, longitude, and elevation | soil taxonomy system[a] | number of replications or blocks | mean annual precipitation | |
| Topography (e.g., flat, rolling hills, glacial moraine), slope % | soil type | site size | characteristic climate descriptor (e.g., temperate moist) | |
| Major Land Resource Area[a] | soil texture | plot size | | |
| Native vegetation | bulk density | treatments | | |
| Drainage | pH | study duration (or start/end dates) | | |
| | soil C, N, P, and K | | | |
| Rainfed or irrigated | water-holding capacity | | | |

| Management | | | | |
|---|---|---|---|---|
| **Tillage** | **Crop system** | **Agricultural amendments** | **Crop management** | **Livestock** |
| Type | crop(s) | type | method | animal species |
| Depth | crop rotation | active ingredient | residue management | animal class |
| Frequency | cropping history | rate | planting data (crop, cultivar) date, rate, depth, method, row spacing, cultivar) | stocking rate |
| Date | cover crop | purpose | cover crop management (crop, cultivar, date, rate, depth, method, row spacing, termination method and date, biomass) | grazing duration |
| Organic yes/no | | application method | yield | |
| | | date | biomass return rate | |

| Sampling | | | | |
|---|---|---|---|---|
| **What** | **When** | **Where** | **How** | **Analysis** |
| plant, soil, air | date | slope position | sample preparation | sampling method |
| | time of day | relative to vegetation | sample storage | brief description or detailed if new or novel |
| | if applicable plant growth stage | depth | | citation |
| | | GPS coordinates | | |

[a] https://catalog.data.gov/dataset/major-land-resource-areas-mlra.

maintained by the U.S. Department of the Interior (https://catalog.data.gov/dataset/major-land-resource-areas-mlra), native vegetation, cropping or land use history, and noting if the site is naturally or artificially drained and if the site receives supplemental irrigation. Soil morphological and developmental forces such as glaciation, alluviation, or sedimentation are also useful as these impact inherent soil properties.

## Soil Characterization

Soil characterization includes providing a taxonomic description of the typical or most common soil(s) within the experimental site. Inclusion of the taxonomic system is needed for international comparisons. Soil type (e.g., Barnes clay loam), soil texture (sand, silt, and clay contents), typical pH, and bulk density are common soil characteristics, preferably by depth increments. Soil chemistry data include soil C (total, organic, and/or inorganic), soil N (total, $NO_3$, and $NH_4$), and other plant-essential nutrients (e.g., P and K). Topographic information can be at a broad-scale site level but can also have much finer granularity related to sampling for response variables. When appropriate, the hillslope position should be noted.

## Experimental Descriptors

Experimental descriptors include experimental design (e.g., randomized complete block, split plot), number of replications or plots, study area, plot size, and treatments.

## Climate and Weather

Climatic data include long-term information such as mean annual temperature and mean annual precipitation. It is also useful to include climate characterization descriptors (Kottek, Grieser, Beck, Rudolf, & Rubel, 2006). Finer granularity of climatic data can also be useful, noting if the region experiences freeze–thaw cycles. Weather data might be as simple as noting the location of the nearest weather recording station, while including a link can be especially useful to the modeling community.

## Management

Management includes soil, crop, and residue management, cropping related amendments, harvest, and animal management. Within the context of an experiment, management parameters might be considered metadata or treatment data.

For example, if the entire study area is managed with a common tillage method or the same cultivar, it would be management metadata. In contrast, if tillage systematically varies across a study based on the selected experimental design, the management is an experimental treatment.

### Tillage

When describing tillage, vague terminology like conventional or conservation should be avoided or if used, these terms need to be clearly described by delineating the implement, mode of soil movement (e.g., inversion tillage), depth, tillage frequency, and the date tillage event(s) occur.

### Cropping System

Cropping systems include listing the crop species with Latin binomial, cultivar, crop rotation, and cropping history. Enough information needs to be provided to understand key parameters if the system is defined as "conventional" or "organic." Clearly indicate if an organic system adheres to the USDA organic standard (https://www.ams.usda.gov/sites/default/files/media/Organic%20Production-Handling%20Standards.pdf) or another organic certification standard when appropriate because it denotes that specific practices have been followed. When cover crops are included in the system, the corresponding information should be included as noted for crop management.

### Amendments

Amendment within this context is used to denote fertilizers, pesticides, or any organic (e.g., animal manure, biochar) or synthetic material applied. The information recommended includes the formulation, active ingredients when applicable, rate, application method, and date applied. For manure, the animal source and nutrient content (C, N, P, etc.) is recommended.

### Crop Management

Crop management includes information about planting rate, row spacing, and when and how cover crops were planted. Likewise, how was the crop was harvested, what was harvested (just grain or a portion of the residue), including other residue management practices (e.g., harvest or burning) should be included. The mass of biomass retained in the field is an important aspect of soil health research because it is the raw material for building nascent soil organic matter, driving nutrient cycling. In those studies that include cover crops, detailed metadata are recommended (species; cultivar; planting date, rate, depth, and method; biomass; termination method and date).

### Livestock Management

Livestock management needs to include animal species, animal class, stocking rate, duration of grazing, and frequency for rotational grazing. Information related to manure applications should be included under agricultural amendments. Manure management may not directly apply to soil health, but if a greenhouse gas mitigation or lifecycle analysis assessment is being conducted, the information would be desirable.

### Sample Collection, Preparation, Storage, and Analysis

Sampling requires its own metadata describing the what, when, where, and how of sample collection. It is the information typically included in the materials and methods section of a paper. *What* describes the sampled material (e.g., soil, plant, air). The resolution of *when* is sample dependent; for example, yield may need only a date or simply a year. Biomass sampling should include a crop-appropriate growth stage description and biomass properties (C, N, or other nutrients). In contrast, if monitoring gas samples, the time scale may be at the sub-hour scale in addition to date and hour. Data describing *where* may include slope position, GPS coordinates, a description of position relative to the crop (e.g., mid- or within-row); soil samples, including for root biomass, should have depth and depth interval recorded. Information related to how the sample was collected includes sampling method, e.g., implement used or if hand collected, whether the sample represents a composite, and frequency of sampling for repeated-measure parameters. Sample handling and preparation documents details such as drying temperature, sieve size, grinding, and storage conditions (e.g., duration, temperature). These details may not be included in a meta-analysis but are important for assuring that like data are being compared and that the data meet quality control standards. For example, oven drying instead of air drying will compromise many biological properties. Another aspect of *how* refers to reporting the methods of analysis. For standard methods, a very brief description and citation may suffice, whereas novel methods should be described with substantially more detail.

## Discussion

The categorial data used in several meta-analyses correspond to the metadata recommended in Table 4.1. For example, a meta-analysis of soil microbial biomass response to C amendments included the following categorical or metadata: microbial biomass method, crop type, cropping system, years of application, type of amendment, soil pH, soil organic content, clay content by depth interval, rate of C input, and rate of N inputs via organic amendments (Kallenbach & Grandy, 2011). Another meta-analysis used soil pH, texture, SOC, and biochar production conditions (feedstock, pyrolysis temperature) and application rate, plus experiment type (field or pot) and study duration to understand soil microbial biomass responses (Zhou et al., 2017). Laganière,

Angers, and Paré (2010) included previous land use, climatic zone, clay content, soil pH, pre-planting disturbance, plantation density, and tree species planted in a meta-analysis assessing C accumulation due to afforestation of former agricultural soils. Mahal, Castellan, and Miguez (2018) conducted a meta-analysis using potential mineralizable N as a soil health indicator and an extensive cadre of metadata including crop type, fertilizer type and rate, cover crop, tillage system, and duration of the experiment in years plus documentation of the incubation method, soil sampling time, soil sampling depth, soil type (texture), pH, bulk density, soil organic matter, SOC, soil total N, $NO_3$ and $NH_4$ concentrations, mean annual temperature and precipitation, crop yield, and state and country where the study was conducted.

## Summary

This chapter emphasizes the importance of collecting metadata that provide adequate information to support the wide range of soil biological, chemical, and physical measurements being quantified for various soil health assessments. Examples of metadata or categorical data that may be useful are suggested but should not be viewed as either extensive or exhaustive. Furthermore, it should not be considered necessary to record every item in the list but rather those data that may contribute to understanding the range of soil health indices being measured for a specific assessment.

## Acknowledgments

The use of trade, firm, or corporation names in this publication is for the information and convenience of the reader. Such use does not constitute an official endorsement or approval by the USDA or the ARS of any product or service to the exclusion of others that may be suitable. The USDA is an equal opportunity provider and employer.

## References

Askari, M. S., & Holden, N. M. (2015). Quantitative soil quality indexing of temperate arable management systems. *Soil and Tillage Research*, 150, 57–67. https://doi.org/10.1016/j.still.2015.01.010

Bünemann, E. K., Bongiorno, G., Bai, Z., Creamer, R. E., De Deyn, G., de Goede, R.,… Brussaard, L. (2018). Soil quality: A critical review. *Soil Biology and Biochemistry*, 120, 105–125. https://doi.org/10.1016/j.soilbio.2018.01.030

Delgado, J. A., Weyers, S., Dell, C., Harmel, D., Kleinman, P., Sistani, K.,… Van Pelt, S. (2016). USDA Agricultural Research Service creates Nutrient Uptake and Outcome Network (NUOnet). *Journal of Soil and Water Conservation*, 71, 147A–148A. https://doi.org/10.2489/jswc.71.6.147A

Del Grosso, S. J., White, J. W., Wilson, G., Vandenberg, B., Karlen, D. L., Follett, R. F., … James, D. (2013). Introducing the GRACEnet/REAP data contribution, discovery, and retrieval system. *Journal of Environmental Quality*, 42, 1274–1280. https://doi.org/10.2134/jeq2013.03.0097

Doran, J. W. (2002). Soil health and global sustainability: Translating science into practice. *Agriculture Ecosystems & Environment*, 88, 119–127. https://doi.org/10.1016/S0167-8809(01)00246-8

Jawson, M. D., Shafer, S. R., Franzluebbers, A. J., Parkin, T. B., & Follett, R. F. (2005). GRACEnet: Greenhouse gas reduction through agricultural carbon enhancement network. *Soil and Tillage Research*, 83, 167–172. https://doi.org/10.1016/j.still.2005.02.015

Kallenbach, C., & Grandy, A. S. (2011). Controls over soil microbial biomass responses to carbon amendments in agricultural systems: A meta-analysis. *Agriculture Ecosystems & Environment*, 144, 241–252. https://doi.org/10.1016/j.agee.2011.08.020

Karlen, D. L., Ditzler, C. A., & Andrews, S. S. (2003). Soil quality: Why and how? *Geoderma*, 114, 145–156. https://doi.org/10.1016/S0016-7061(03)00039-9

Kottek, M., Grieser, J., Beck, C., Rudolf, B., & Rubel, F. (2006). World map of the Köppen–Geiger climate classification updated. *Meteorologische Zeitschrift*, 15, 259–263. https://doi.org/10.1127/0941-2948/2006/0130

Laganière, J., Angers, D. A., & Paré, D. (2010). Carbon accumulation in agricultural soils after afforestation: A meta-analysis. *Global Change Biology*, 16, 439–453. https://doi.org/10.1111/j.1365-2486.2009.01930.x

Liebig, M., Varvel, G., & Honeycutt, W. (2010). Guidelines for site description and soil sampling, processing, analysis, and archiving. In R. Follett (Ed.), *GRACEnet Sampling Protocols* (pp. 1–5). Washington, DC: USDA–ARS.

Liebig, M. A., Franzluebbers, A. J., Alvarez, C., Chiesa, T. D., Lewczuk, N., Piñeiro, G., … Sawchik, J. (2016). MAGGnet: An international network to foster mitigation of agricultural greenhouse gases. *Carbon Management*, 7, 243–248. https://doi.org/10.1080/17583004.2016.1180586

Liebig, M. A., Franzluebbers, A. J., & Follett, R. F. (Eds.). (2012). *Managing agricultural greenhouse gases: Coordinated agricultural research through GRACEnet to address our changing climate*. Waltham, MA: Academic Press.

Mahal, N. K., Castellano, M. J., & Miguez, F. E. (2018). Conservation agriculture practices increase potentially mineralizable nitrogen: A meta-analysis. *Soil Science Society of America Journal*, 82, 1270–1278. https://doi.org/10.2136/sssaj2017.07.0245

Robinson, D. A., Panagos, P., Borrelli, P., Jones, A., Montanarella, L., Tye, A., & Obst, C. G. (2017). Soil natural capital in Europe; A framework for state and change assessment. *Science Reports*, 7, 6706. https://doi.org/10.1038/s41598-017-06819-3

Zhou, H., Zhang, D., Wang, P., Liu, X., Cheng, K., Li, L., … Pan, G. (2017). Changes in microbial biomass and the metabolic quotient with biochar addition to agricultural soils: A meta-analysis. *Agriculture Ecosystems & Environment*, 239, 80–89. https://doi.org/10.1016/j.agee.2017.01.006

# 5

# Soil Health Assessment of Agricultural Lands

*Diane E. Stott, Brian Wienhold, Harold van Es, and Jeffrey E. Herrick*

## Summary

Multiple approaches for assessing soil health have been developed, tested, promoted, and evaluated during the past three decades. This includes scorecards, farmer test kits, and various indexes including the Soil Conditioning Index (SCI), Rangeland and Forestland assessment, AgroEcosystem Performance Assessment Tool (AEPAT), Soil Management Assessment Framework (SMAF), Comprehensive Assessment of Soil Health (CASH), Land-Potential Knowledge System (LandPKS), and most recently the Soil Health Assessment Protocol and Evaluation (SHAPE). This chapter provides a brief overview of soil health assessment and briefly summarizes the development of those indexes.

## Overview of Assessments

There is a broad public recognition that soil is a finite resource and conserving it is essential. This has led to an increasing number of public and private efforts engaged in evaluating soil health not just in the United States of America (USA), but also around the world. Driven by an increased recognition that anthropogenic factors are affecting the quality and functioning of soils, assessments are being used to quantify impacts of cropping systems and soil management practices on carbon (C) flows and cycles, nutrient cycling, erosion, and greenhouse gas (GHG) emissions. For anyone interested in establishing a

*Soil Health Series: Volume 1 Approaches to Soil Health Analysis*, First Edition.
Edited by Douglas L. Karlen, Diane E. Stott, and Maysoon M. Mikha.

quantitative soil health baseline using laboratory measurements, we suggest reviewing sampling and laboratory protocols discussed in Chapters 1 and 2 of Volume 2. Thereafter, resampling is recommended at time intervals appropriate for detecting significant changes in soil health for the question being asked (e.g., productivity, erosion, carbon storage, leaching loss, GHG emissions). The time interval will depend primarily on anticipated rates of change, indicator sensitivity, and sampling intensity. Typically, this interval will be every 3 to 5 yr for humid and sub-humid climates and 5 to 10 yr under arid or semiarid, non-irrigated regions (Stott, 2019). For irrigated areas, many indicators are likely to change more rapidly and may need to be resampled more frequently. Once a trend line has been established, soil health sampling should be based on producer or organization goals and guided by the question(s) being asked (e.g., productivity, environmental impact, potential soil and crop management changes). Furthermore, from the beginning, critical meta-data (Vol. 1, Chapter 4) should also be recorded with the measurements in a long-lasting format.

After sampling indicators from the same field, farm, or other designated area over time, the data record can be used to evaluate soil health with or without a specific assessment tool, statistical method, or pre-defined protocol for interpretation. However, this may take several years to be of value to producers or organizations, because both spatial and temporal variability of soils will likely require sampling for many years to establish a meaningful baseline. Therefore, the use of an assessment framework is a proactive process for interpreting indicator measurements capitalizing on science-based soil research conducted over the last several decades or more.

An assessment framework facilitates comparison of results from similar soils and production environments to determine where a field may be on a soil health continuum. Such tools need to be flexible regarding selection of soil functions and indicators to ensure the measurement and ensuing assessment is meaningful and suitable for agroecosystem and overall management goals. To create robust interpretations, calibration databases must be developed by collecting and analyzing soil samples from many different ecosystems as well as soil and crop management systems using comparable sampling and analytical methods (see Chapters 1 and 2 of Volume 2). The databases can be built using different types of field observations (Vol. 1, Chapter 8) and critical meta-data (Vol. 1, Chapter 4) for each. Descriptions of several assessment frameworks are available in the literature, although many were developed for localized areas or specific applications (e.g., forest or contaminant evaluations, as discussed in Vol. 1, Chapters 6 and 7). A summary of potential soil health assessment frameworks developed for various regions is presented in the following subsections.

# Assessment Frameworks

## Initial Soil Health Frameworks

Karlen and Stott (1994) described the first general procedure for designing a site-specific framework using soil biological, chemical, and physical characteristics. Building on principles of systems engineering (Wymore, 1993), they identified critical soil health functions that assumed high-quality agricultural soils would: (i) allow rain or irrigation water to enter the soil surface (infiltration); (ii) absorb, retain, and release water to meet plant evapotranspiration demand; (iii) resist degradation by wind and water erosion; (iv) mitigate contaminants, and (v) sustain plant growth.

The primary soil health indicator for quantifying the water entry function is the infiltration rate, which has secondary effects reflected by surface crusting, surface roughness, soil macroporosity, and crop residue cover. To facilitate water transfer and adsorption, hydraulic conductivity is the primary indicator although soil water relations are also influenced by texture, capillary water content, bulk density, porosity, and bio-pores created by plant roots and macro-organisms such as earthworms. With regard to soil resistance to degradation, aggregate stability is the primary soil health indicator, which is in turn influenced by soil mineralogy, quantity and type of cations and carbohydrate within the soil, microbial biomass, shear strength, and heat transfer capacity. Finally, for sustaining plant growth, properties that are important include rooting depth (as affected by restrictive layers, inherent profile depth, and soil texture), water relations (i.e., plant-available water holding capacity, drainage characteristics, and organic C content), essential nutrient availability (as affected by pH, organic C, and micro- and macronutrient concentration), salinity, heavy metals, and organic contaminants.

After choosing applicable indicators, various weights can be assigned to the primary indicators, with the total adding to 1.0. Following Wymore's (1993) protocol, secondary indictors, influencing or contributing to each primary indicator can also assigned weights that also collectively add to 1.0. This approach thus creates a hierarchical index (i.e., framework) with as many levels as are needed to quantitatively describe and evaluate the soil health question.

As discussed in numerous publications describing initial soil health assessment framework development, Wymore (1993) provides three basic response (i.e., scoring) curves. They are: (i) more-is-better (upper asymptotic sigmoid curve), (ii) less-is-better (lower asymptotic sigmoid curve), and (iii) midpoint optima (Gaussian function) as shown in Figure 5.1 (Karlen & Stott, 1994; Andrews et al., 2004). Again, as stated above, to develop accurate, meaningful, and science-based response curves, data from many locations, soil types and conditions, climatic regimes (precipitation and temperature), and management practices are required.

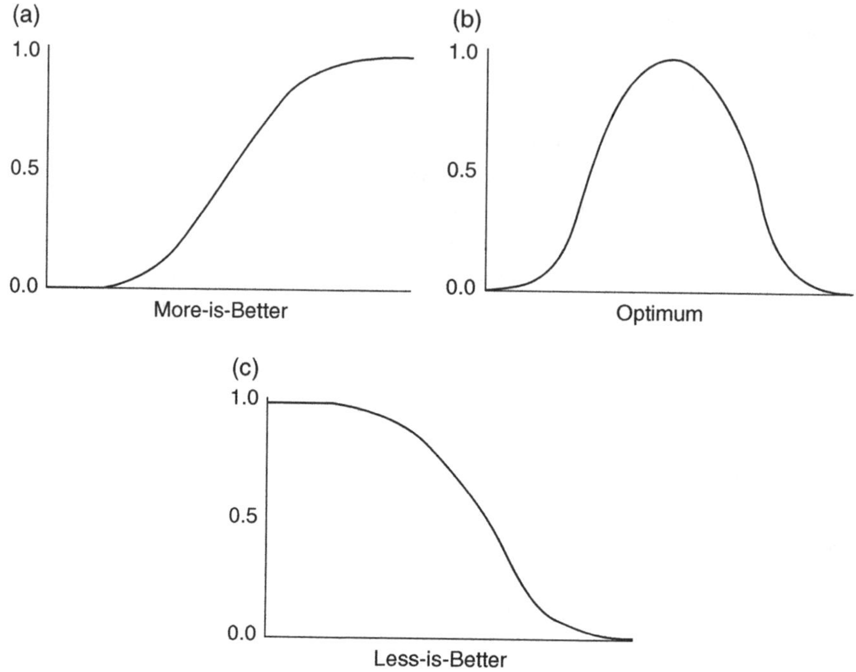


**Figure 5.1** General shapes for standard scoring functions. From left to right: more-is-better response curve (a), an optimum range response curve (b) and a less-is-better response curve (c) adapted from Karlen and Stott (1994).

## Soil Health Cards

Parallel with development of soil health assessment frameworks was creation of soil health cards, with one of the first being published by Romig et al. (1995). Their approach was built on recommendations from Arshad and Coen (1992) which included having qualitative (descriptive) information included in soil quality monitoring programs. This includes visible signs of wind and/or water erosion, poor drainage (as evidenced by standing water), surface crusting or sealing, soil color, texture, structure, or consistence, and plant-based observations such as poor or patchy stand establishment, coupled with soil organic matter, aggregate stability, salinity, pH and nutrient availability measurements. Soil scorecard development using those criteria was also consistent with perspectives of farmers in the Palouse region of Washington and Idaho where they had been experimenting with low input and other alternative farming practices (Granatstein, 1990). Those farmer groups typically used one of three approaches: (i) crop rotations to help build soil, reduce erosion, and in some cases decrease input costs; (ii) significant reductions in purchased inputs; or (iii) use of alternative or biological products as major components of their production systems (Harris & Bezdicek, 1994).

Soil health scorecards have been available for 30 yr through state USDA-NRCS offices (https://www.nrcs.usda.gov/wps/portal/nrcs/detailfull/soils/health/assessment/?cid=nrcs142p2_053871) and University Extension programs that have tailored the cards for their regions. Many countries around the world have used soil health scorecards with varying success and reasons for developing and implementing the cards. This includes soil erosion (e.g., Canada and the USA), soil fertility (e.g., Kenya, Tanzania and South Africa), soil contamination (e.g., Netherlands and China), condition of forest lands (e.g., Austria and the USA), and land degradation due to salinization (e.g., Australia and New Zealand). The Agriculture and Horticulture Development Board (AHDB) in the United Kingdom (UK) developed a "GREATsoils" initiative that includes technical information and research on soil-related issues. This Soil Health Partnership is one AHDB example designed to increase overall understanding of soil biology and thus measure and manage soil health (https://ahdb.org.uk/greatsoils) using visual, field observations and local knowledge regarding visual responses associated with soil degradation. Similarly, India launched their Soil Health Card Scheme in early 2015 (https://soilhealth.dac.gov.in/) with the Department of Agriculture & Cooperation under the Ministry of Agriculture and Farmers' Welfare.

Scorecard observations usually include estimates of surface cover (mulch and canopy); soil structure (based on crusting and sealing, penetrometer resistance, surface layer color, and signs of erosion that are then verified by laboratory measurements of water aggregate stability, organic matter content, and bulk density); water infiltration rate (e.g., ponding); biodiversity (presence of earthworms, spiders, and insects); and plant growth (seedling emergence, lack of drought stress, and profile root development and distribution). Many scorecards also include a nutrient analysis report as part of an overall assessment.

Soil health scorecards are generally well received by farmers because their format is simple with tables or bar charts. By design, they summarize indicator measurements in ways similar to what they are accounstomed to from their soil fertility analyses (Andrews et al., 2003). However, with a little training in interpretation, soil health scorecards can be enhanced using other formats (e.g., spiderwebs) as discussed by Wander et al. (2002).

## Soil Conditioning Index (SCI)

The SCI was developed by the USDA-NRCS and adopted to evaluate cropland management practices in the United States. It predicts the consequences of soil and crop management actions (e.g., tillage type and intensity or crop residue removal) on soil organic carbon (SOC) and has been embedded within the Revised

Universal Soil Loss Equation, Version 2.0 (RUSLE2) (Zobeck et al., 2007). The SCI predicts qualitative SOC changes in the top 10 cm (4 inches) of soil based on: (i) the amount of organic material, both plant and animal, returned to the soil, (ii) number and type of field operations that result in the breakdown and enhancement of organic materials, and (iii) annual rates of wind and water erosion. The three components are weighted 40, 40, and 20%, respectively. The SCI values can be either negative or positive, with positive numbers considered to be conservation and soil health management systems. Several USDA-NRCS practice standards, such as Conservation Crop Rotation (328) and Residue and Tillage Management, No Till (329), require a positive SCI rating to fulfill contract criteria for conservation enhancement programs such as the Conservation Security Program (CSP) (USDA–NRCS, 2014).

## AgroEcosystem Performance Assessment Tool (AEPAT)

Most soil health assessment frameworks utilize spreadsheets, but AEPAT is a software tool developed to assess relative soil management effects on multiple indicators of agricultural sustainability, including soil health (Liebig et al., 2004). AEPAT can be downloaded (from https://data.nal.usda.gov/dataset/agroecosystem-performance-assessment-tool) and allows users to choose which agroecosystem functions are of interest and most meaningful to their questions. It then assigns potential indictors and weights for each calculation, in a way similar to that laid out by Karlen and Stott (1994). Users select how each indicator is related to various specific soil functions, using more is better, less is better or optimal values. They can also choose the type or shape of each scoring function (i.e., linear, exponential, logarithmic, or sigmoidal) by entering descriptive values based on the user's data. Ultimately, AEPAT calculates performance scores for each indicator and function to provide an overall agroecosystem performance score between 0 and 100. The AEPAT was very effective for the Great Plains farming systems comparisons made by Liebig et al. (2004), but it required so many inputs regarding response curve shape and boundaries, that it never moved beyond being used as a research tool.

## Rangeland and Forestland Assessment

None of the above assessments were developed for soil health assessment of rangeland or forestland. The technical reference, "Interpreting Indicators of Rangeland Health" or IIRH (Pellant et al., 2020), incorporates some in-field soil health indicators as does the forest based "Soil Vital Signs: A New Soil Quality Index (SQI) for Assessing Forest Soil Health" (Amacher et al., 2007). The initial published version of IIRH ("Version 3") became available in 1999. The most recent (Version 5) retains all of the original in-field soil health indicators including compaction, soil aggregate stability, and soil surface loss and degradation, as well as several soil surface attributes. The IIRH represented a significant departure from traditional rangeland assessment protocols that were almost entirely based on vegetation cover and species composition. The assessment protocol is similar to the cropland-focused tools in that it requires users to evaluate each indicator against its natural potential, which is a function of climate, topography, and relatively inherent or static soil properties including texture, profile depth, and mineralogy (Pyke et al., 2002). "Ecological sites" are used to group soils for assessment, and a simple "reference sheet" describing the potential for each indicator is available for each ecological site. The IIRH protocol has been adopted by both the NRCS and Bureau of Land Management, as well as internationally which has resulted in an extraordinarily large, standardized, soil health indicator database (Herrick et al., 2010; McCord et al., 2020). The IIRH has also led to the training of thousands of individuals in making soil health observations, many of whom had little or no previous exposure to soils. For additional information on the assessment of forest soil health, we refer you to Volume 1, Chapter 6.

## Soil Management Assessment Framework (SMAF)

The SMAF, developed by the USDA-ARS and USDA-NRCS, provides site-specific interpretations for several soil health indicators, primarily for agriculturally managed lands. It uses measured data to assess management impacts on soil function via three steps: indicator selection, indicator interpretation, and if desired integration into an index (Andrews et al., 2004).

Soil taxonomic groups provide the foundation for the SMAF assessments. The algorithms, used to score the indicator values, are modified based on essential soil suborder characteristics that provide a contextual basis for indicator interpretation. Besides soil taxonomic information, landscape characteristics, environmental influences (e.g., climate), and human values (e.g., land use, management goals, and environmental protection) are considered within the SMAF.

The SMAF includes soil physical, chemical, and biological characteristics that are dynamic and management sensitive. Currently, the SMAF has scoring curves or interpretative algorithms for 13 indicators. These are: wet macroaggregate stability (WAS), bulk density (BD), water-filled pore space (WFPS), available water capacity (AWC), electrical conductivity (EC, salinity), pH, sodium adsorption ratio (SAR, used only in naturally high Na soils, western irrigated lands, and specialized situations such as high-tunnels used primarily for vegetable crop production), extractable P and K, soil organic carbon (SOC), microbial biomass C (MBC), potentially mineralizable N (PMN) and β-glucosidase activity (BG), (Andrews et al., 2004; Wienhold et al., 2009; Stott et al., 2010). Five of those indicators (WAS, K, SOC, MBC, and BG) use the more-is-better response curve, (Figure 5.1), BD, EC and SAR are represented by the less-is-better curve, while the remainder are described using an optimal response curve. The scored indicators can be summed to create an overall soil quality index (SQI) (Andrews et al., 2004) or divided into physical (WAS, BD, WAC, WFPS), chemical (pH, EC, P, K, SAR), and biological/biochemical (SOC, MBC, PMN, BG) components. It is generally recommended to measure at least two indicators from each sector to ensure an accurate, meaningful assessment.

To increase sensitivity of the SMAF and further its flexibility for assessing other soil and crop management impacts, additional indicator scoring curves should be developed. This can be done through a multistep process that begins with indicator selection, then determining the type of relationship between the new indicator and specific soil functions, identifying appropriate algorithm(s) that describe that relationship, and ultimately validating the new scoring curve(s) (Andrews et al., 2004; Wienhold et al., 2009).

Three papers compared the SMAF with the SCI in Iowa (Karlen et al., 2008), Western Texas (Zobeck et al., 2008), and Oklahoma (Zobeck et al., 2015). Each found the SMAF and SCI to be positively correlated and provide similar assessment results, but the SMAF provided more resolution when evaluating management impacts on soil health within agroecosystems. When the SCI-SOC values were compared with the SMAF-SOC values (a more direct comparison than the full suite of SMAF indicators), the SMAF-SOC values were more successful in detecting subtle differences among the cropping systems. However, the SCI provides information regarding soil erosion that is not yet available through a SMAF analysis. AEPAT and the SMAF also showed general agreement (Wienhold et al., 2006), but input requirements and intended uses for the two tools are different, making a direct comparisons difficult.

The SMAF has been used in several U.S. regions and internationally. Within the North American Great Plains, soil quality effects of conventional and alternatively managed cropping systems were compared at eight sites located near Swift Current, SK, Canada; Sidney, MT; Mandan, ND; Fargo, ND; Brookings, CO; Akron, CO; Mead, NE; and Bushland, TX (Wienhold et al., 2006). At each site, every treatment had been in place for more than 10 yr. Alternative management strategies included reduced tillage intensity and a decreased incidence of fallow compared to conventional practices. Soil sampling was conducted three times each growing season from 1999 through 2002. Samples were analyzed for physical, chemical, and biological indicators of soil health. A major finding was that temporal variation within several measured indicators made dynamic assessment essential. The results showed that WAS, WFPS, MBC, and particulate organic matter (currently not a SMAF indicator) were sensitive to management and that overall soil health improved with alternative management practices.

The SMAF was also used to assess soil health in several Conservation Effects Assessment Project (CEAP) watersheds. Within the South Fork of the Iowa River

Watershed, fifty corn (*Zea mays* L.) fields with poorly performing sections were identified (Stott et al., 2011). There were no significant differences between the two performance zones when analyzed collectively, but by using SMAF indicator scores, SOC, BD EC, and MBC were significantly lower in poor canopy areas; however, no single indicator was consistently low across all 50 fields. Most fields had multiple SMAF indicators with significantly lower (>0.10 difference) ratings in poor areas than in corresponding areas with well-developed plant canopies. Soil health assessment on a field-by-field basis thus provided an approach for identifying specific soil-based measurements associated with poor canopy development.

Stott et al. (2013) summarized soil health response within five small Texas Vertisol watersheds: a hayed native prairie (NP), two Coastal Bermuda grass (Cynodon dactylon) sites that were hayed for 57 (CB) and 31 (CBTL) years, respectively, and two sites that were continuously cropped for 57 (RC) and 31 (RCTL) years. The CBTL and RCTL fields also had turkey (*Meleagris gallopavo*) litter amendments. The overall SMAF SQI ranged from 75 to 94% in the following order: RC ≈ CB < RCTL ≈ NP < CBTL. When separated into sectors, all systems had nearly the same chemical rating, 94 to 95%, physical ratings range from 66 to 84%, RCTL ≈ CB ≈ RC < NP ≈ CBTL. The nutrient attributes were significantly higher (99%) in the RC, RCTL, and CBTL, with NP being significantly lower (47%). This highlighted the impact of haying NP with no nutrient returns. Biological and biochemical soil health ratings ranged from 44 to 81%, with the NP performing significantly better (81%) and RC significantly lower (44%) than the other management systems. Overall, turkey litter amendments improved soil health, demonstrating that carefully managed Bermuda grass pastures can be more productive than managed native grass pastures, primarily because of improved plant nutrient availability.

Karlen et al. (2014) measured ten soil health indicators in five Midwestern USDA-ARS experimental watersheds: South Fork of the Iowa River watershed in north-central Iowa; Walnut Creek watershed near Ames in central Iowa; Cedar Creek watershed, part of the larger St. Joseph River watershed in northeastern Indiana; Mark Twain watershed is part of the Salt River Basin in northeastern Missouri; and Upper Big Walnut Creek watershed in central Ohio. Using SMAF scoring algorithms to normalize and score the data, significant differences among the five watersheds were found in all but the extractable P indicator. This study documented that watershed-scale monitoring of soil health is feasible and can provide useful information for conservation planning.

Within the Salt River Basin, located in the Central Claypan Region of Missouri, multiple soil health indicators were measured within 15 different annual and perennial cropping systems found in that region (Veum et al., 2015). Ten biological, physical, chemical, and nutrient indicators were measured and scored with the SMAF. Biological and physical indicators were most sensitive to management, especially within the top 5 cm layer. Biological indicators under the diversified no-till system with cover crops was 11% greater than for no-till without cover crops, and 20% greater than for mulch-till without cover crops. Effects of crop rotation were most evident for PMN with a 64% lower score following corn than soybean (Glycine max (L.) Merr.). Soil nutrient ratings were also significantly affected by biomass removal. Benefits of conservation went beyond reduced soil erosion and improved water quality as highlighted by enhanced soil health potential due to improved biological soil functions. Based on the assessment, implementing conservation practices on marginal and/or degraded soils in the claypan region can enhance sustainability for both annual cropping systems and grasslands.

Hammac et al. (2016) assessed soil health within the pothole-dominated Cedar Creek sub-watershed which contributes to the St. Joseph River watershed and drains into the Western Lake Erie Basin in northeastern Indiana. Soil samples were collected at three landscape positions (summit, mid-slope, or toe-slope), from conventional and no-till fields planted to corn, soybean or perennial grass, before measuring ten soil health indicators and using SMAF to normalize and score the data. Surface (0–15 cm) physical, chemical, and nutrient component indices were high, averaging 90, 93, and 98% of the optimum, respectively. The biological component had the lowest score, averaging 69% of the optimum. They concluded crop selection had a greater impact on soil health than tillage treatment, but reanalysis of the management metadata revealed that almost all sampled no-till fields were being managed using rotational tillage. Indicator comparisons indicated perennial grass sites generally had higher values than either corn or soybean fields. The soybean phase of the rotation also tended to score higher than the corn phase. Uncultivated perennial grass sites had higher overall SQI values, as well as higher physical, chemical, and biological component values than rotational no-till or chisel-disk systems. Except for a few physical indicators, chisel-disk and rotational no-till generally showed no significant differences in component or overall SQI values. Regarding landscape position, toe- and mid-slope positions had higher physical, biological and overall SQI values than summit positions. This study also showed that when traditional statistical approaches were used for this data, few significant differences due to soil type slope position were detected, whereas normalizing the data with the SMAF allowed differences to be seen.

The SMAF was used to assess pasture management effects on both soil and runoff water quality in Arkansas (Amorim et al., 2020a). That study was conducted over 15 yr using small watersheds (0.14 ha) where Bermuda grass was the major forage species. The management practices included harvesting grass for hay, continuous grazing, rotational grazing, rotational grazing with an unfertilized but grazed buffer strip, and rotational grazing with an unfertilized, fenced and therefore ungrazed buffer strip. Soil samples were collected at three topographic positions within the watershed and in the buffer strip when present. Runoff was collected and analyzed for total P, total dissolved C, and total suspended solids in 2017. Continuous and rotational grazing had the highest SMAF SQI values which did not differ from the fenced, ungrazed, unfertilized buffer strip. Differences in SQI values among the management practices were due to differences in pH, EC, P, and K. The SMAF SQI also explained 34 and 28% of the variation in total P and total dissolved C loads in runoff. The SMAF SQI values were useful for identifying long-term pasture management practice effects developing conservation pasture management practices that can enhance soil and water quality.

The effect of cover crops or poultry litter (biocover) on soil quality were compared across three cropping systems in Arkansas using SMAF (Amorim et al., 2020b). The treatments were part of a 15-yr cropping system study with corn, cotton (*Gossypium hirsutum* L.), or soybean based with biocover treatments applied following harvest of the primary crops. Soil indicators included: pH, SOC, BD, P, K, EC, and SAR. The SMAF SQI was greater for corn and cotton rotations than for soybean rotations. The poultry litter biocover resulted in the highest SOC, pH, K, and BD scores, but the lowest SQI. The low overall SQI reflected a low SMAF P score because of the high soil-test P content and potential for decreasing water quality due to high dissolved P concentrations in runoff from the poultry litter treatment.

The SMAF has also been used for soil health assessments in several countries around the world. For example, in the south-central region of Brazil, the SMAF was used to evaluate sugarcane (*Saccharum officinarum* L.) expansion impact on soil health (Cherubin et al., 2016). Despite having been developed primarily for the USA, the framework was sensitive enough to detect soil health differences due to land use change (i.e., native vegetation-pasture-sugarcane). A resin-extractable P scoring curve had to be developed for this region, but overall, the eight soil health indicators scored using existing SMAF algorithms were sufficiently robust to detect significant differences in these tropical areas. Transition from native vegetation to extensive pasture used for grazing significantly decreased soil chemical, physical, and biological indicator values. Based on overall SQI values, soils under native vegetation were functioning at 87% of their potential capacity, while pasture soils were functioning at only 70%. Conversion of long-term pasture to sugarcane resulted in slight SQI improvements (i.e., 74% of potential capacity), primarily because of improved soil fertility. Cherubin et al. (2017) used data from five southern Brazil studies to evaluate the SMAF's potential for assessing diverse land-use and management practices on soil health. The datasets included horizontal and vertical distribution of soil properties in: (i) a long-term orange (*Citrus sinensis*) orchard; (ii) sites converted from native vegetation to agricultural crops; (iii) cassava (*Manihot esculenta*) sites with a history of short-term tillage; (iv) sites comparing mineral fertilizer and pig slurry applications with various tillage practices; and (v) row versus inter-row effects within a long-term, no-tillage crop production site. The soils were classified as Oxisols with clay contents ranging from 180 to 800 g $kg^{-1}$. Six soil health indicators (pH, P, K, BD, SOC and MBC) were measured, scored using the SMAF curves, grouped and summed to provide chemical, physical, and biological component values, and combined into an overall SQI. The SMAF algorithms logically and accurately transformed the indicator values into unitless scores ranging from 0 to 1, thus enabling indicators with different measurements to be combined. Overall, the study confirmed that SMAF could be used as an effective tool for assessing soil health in Brazilian soils, thus helping farmers, land managers, and policymakers choose actions that will ultimately result in improved and sustainable land-use.

Da Luz et al. (2019) evaluated the impacts of several land uses (native vegetation, pasture, sugarcane, no-tillage, and integrated crop-livestock systems) on soil quality in three agricultural regions of Parana state, southern Brazil. SMAF scores detected that long-term conversion from native vegetation to agricultural land uses (i.e., pasture, no-tillage/integrated crop-livestock system, or sugarcane) reduced soil quality. Nevertheless, conservation systems, e.g., the no-tillage, associated or not with the integrated crop-livestock system are promising alternative to enhance soil quality by increasing C content and soil chemical fertility compared to degraded pasture or conventional sugarcane cultivation. The authors also warned that soil physical changes should be monitored to alleviate soil compaction in no-tillage cropping systems in Brazilian tropical soils. In another Brazilian study, Lisboa et al. (2019) used SMAF to assess the short-term effects of sugarcane residue management in Brazil. This study was conducted at two sites with different soils (Oxisol vs. Ultisol) for 2 yr. Residue management treatments were 0, 50, or 100% removal. The study concluded that 100% removal at the Oxisol site resulted in a decline in soil physical properties while removing 0 or 50% resulted in an improvement in soil quality compared to 0% removal. Residue management at the sandy loam Ultisol site did not influence soil quality. This study demonstrated the site (soil) specific nature of management impacts and that SMAF could detect changes in soil quality over relatively short time periods. Ruiz et al. (2020) measured soil health in 3- and 7-yr-old recently constructed soils (i.e., Technosols) used for restoring post-mining landscapes. The Technosols were constructed with limestone spoil in São Paulo state, Brazil and were compared to an adjacent natural soil (NS; Rhodic Lixisol) under pasture. The SMAF was used to detect soil health changes during reclamation. After 3 and 7 yr under sugarcane cultivation, the Technosols showed similar SQI indexes (0.70 and 0.67, respectively) to that of the native soil (0.69), while the pasture was significantly higher (0.88). The Technosols recovered most of the ecosystem services expected for healthy soils.

In northeast Spain, the SMAF was used to assess soil health changes 3 yr after transitioning from dryland to irrigated crop production (Apesteguía et al., 2017). The field experiment included long-term conventional tillage (CT), minimum tillage (MT) and no-tillage (NT) treatments with either barley (*Hordeum vulgare* L.) or wheat (*Triticum aestivum* L.) crops. To assess soil health, BD, WSA, AWHC, available P, extractable K, pH, EC, SOC and MBC indicators were measured. The SMAF analysis separated the treatments enabling them to be ranked and used as guidelines for making conservation decisions. An important finding from this study was that the algorithm used to assess BD needed to be recalibrated for these sandy Mediterranean soils. This provides an example of how the SMAF has been continuously refined since its inception.

The SMAF was also shown to be a robust tool for assessing land preparation systems on Vertisols in the Central Highlands of Ethiopia (Erkossa et al., 2007). In South Africa, the SMAF was successfully used to evaluate conservation management strategies being used by subsistence farmers in the Eastern Cape (Gura & Mnkeni, 2019a, 2019b). The assessment showed that retaining crop residues in a maize-wheat-soybean rotation resulted in the highest SQI values. Meanwhile, in the Cape region of South Africa, Swanepoel et al. (2015) used SMAF to evaluate the soil tillage effects on soil health of kikuyu (*Pennisetum clandestinum*)–ryegrass (*Lolium* spp.) pasture systems. Based on SMAF scores they documented that extended use of tillage in dairy forage systems had a profound negative impact on physical and biological soil health indicators.

## Comprehensive Assessment of Soil Health (CASH)

The CASH framework, initially known as the Cornell Assessment of Soil Health (Idowu et al., 2008; Moebius-Clune et al., 2016) was developed at Cornell University and designed to offer a comprehensive set of soil measurements to support soil and crop management decisions and associated applied research (https://soilhealth.cals.cornell.edu/). It provides an integrated soil health evaluation, but the primary goal is to identify specific constraints and link them to on-farm management solutions (i.e., reduced tillage, cover cropping, crop rotations, organic amendments). The CASH scores rank relative soil functioning with respect to crop production and environmental impact and are expressed as percentile ratings for comparing measured values to a known population distribution for each textural group. The CASH sampling protocol includes a representative minimally disturbed soil sample and field measurements of penetration resistance. The CASH was created using several SMAF concepts, but shifted some indicators to provide greater sensitivity and faster processing (i.e., substituting autoclave-citrate extractable [ACE] proteins for PMN or permanganate-oxidizable carbon [POXC] for MBC). Recently, random forest regression was used to estimate available water content (AWC) from other indicators rather than making direct measurements to streamline the assessment.

The CASH has been applied to: (i) quantify soil health impacts of land degradation after conversion from forest to cropland in Kenya (Moebius-Clune et al., 2011), (ii) assess effects of farmyard manure application in Pakistan (Iqbal et al., 2012), (iii) evaluate soil and crop management impacts on soil health in New York (Idowu et al., 2009; Nunes et al., 2018), (iv) evaluate crop and landscape effects in India (Frost et al., 2019), (v) coffee culture in Colombia (Rekik et al., 2018), and (vi) tillage and organic management practices in North Carolina (van Es & Karlen, 2019). Those studies generally showed that indicators associated with labile forms of carbon and nitrogen, especially POXC, ACE protein and WAS, were most sensitive to management effects and therefore, good indicators of soil biological and physical health. Furthermore, van Es and Karlen (2019) in a re-analysis of published data concluded that tillage-induced corn yield differences were most strongly correlated (in order of decreasing $R^2$) to ACE protein, active C (AC), Mn, respiration (RESP), and WAS, while soybean yields were most closely associated with AC, RESP, Mg, Mn, and ACE protein. For both crops, the correlation between yield and those soil health indicators was better than between yield and total SOM.

Finally, several nationwide meta-analysis studies have confirmed the sensitivity of soil biological, physical, and chemical properties, used in CASH and SMAF, to anthropogenic and inherent factors. For example, Nunes et al. (2020a, 2020b, 2020c), using data from 302 studies conducted across the continental USA, showed that several biological (i.e., SOC, MBC, microbial biomass-N, RSP, AC, ACE protein, BG), physical (i.e., WAS, BD, penetration resistance) and chemical (pH and nutrient) soil health indicators are sensitive to management practices (i.e., tillage intensity). However, those soil properties' response to management and land use was site-specific and reflected soil type, texture, climatic, study duration, sampling depth, and cropping systems. Overall, those studies highlighted the importance of site-specificity when interpreting soil property measurements to assess anthropogenic effects on soil health.

## Land-Potential Knowledge System (LandPKS)

The LandPKS (https://LandPotential.org) is a global platform designed to easily identify soils and obtain land-related knowledge and information through a mobile app (Herrick et al., 2013). It also allows virtually anyone, regardless of their knowledge of soils, to assess and monitor soil and vegetation with a suite of input modules. LandPKS supports knowledge and information generation and sharing needed to sustainably support increased agricultural production and other ecosystem services (Herrick et al., 2019).

One of the most useful features of LandPKS is the integration of soil characterization with soil health assessment and monitoring. This ensures users will have the information they need to compare their local soil health values with those from similar soils. The app includes a soil texture key, and each step for its use is explained by a short GIF-type video or graphic illustration. Soil color is determined using a camera linked to a standard reference such as a "grey card" or a 3M canary yellow post-it-note. A cloud-based algorithm ranks the soils mapped in the area based on their similarity to the user's inputs. Another set of on-phone algorithms predicts plant-available water holding capacity and infiltration capacity as a function of soil texture and user-estimated soil organic matter content. The USDA Land Capability Class is predicted based on texture, slope, and several other user inputs (Quandt et al., 2020).

As of 2020, the LandPKS Soil Health module provides a simple way to document the status of the USDA-NRCS in-field cropland indicators. Expanding a simple "yes/no" system to multiple classes for each indicator addressed user concerns that the framework didn't account for differences in potential (i.e., "good" aggregate stability for a sand soil in an arid climate is much lower than for loamy soil in a humid climate). In response to users, several other indicators related to soil erosion have been incorporated into LandPKS enabling the system to also be used for monitoring.

## Evolving Soil Health Assessment Activities

Karlen et al. (2019) documented how the soil health concept evolved over the past several decades and emphasized the need to scientifically advance monitoring and assessment protocols by (i) improving indicator scoring tools, (ii) developing national monitoring protocols, and (iii) identifying new soil health indicators. To meet those needs, the USDA-NRCS and USDA-ARS initiated a meta-analysis project focused on indicator interpretation and tool development. A national database was compiled using soil health data from 456 published studies and from Cornell Soil Health Laboratory. Thereafter, the database was used to develop a new tool, identified as Soil Health Assessment Protocol and Evaluation, or SHAPE (Nunes et al., 2021). SHAPE builds on conceptual frameworks established by SMAF and CASH protocols being a comprehensive Bayesian linear regression model developed to improve scoring for SOM (Nunes et al., 2021). This new tool scores indicator measurements using a conditional cumulative distribution function that incorporates both categorical factors (five soil suborder classes and five texture classes) and continuous climate variables (mean annual precipitation and mean annual temperature), thus embracing both the SMAF and CASH approaches to create the next generation soil health assessment tool. Reanalysis of published case studies confirmed sensitivity of SHAPE scoring curves to land use and field-scale management (Nunes et al., 2021). Full development of the SHAPE will help meet the growing demand for an accessible, interpretive, and quantitative scoring curve that provides regionally relevant knowledge regarding the status of soils in response to various agronomic and conservation initiatives.

## References

Amacher, M. C., O'Neil, K. P., & Perry, C. H. (2007) Soil vital signs: A new soil quality index (SQI) for assessing forest soil health. Research Paper RMRS-RP-65WWW. Fort Collins, CO: USDA, Forest Service, Rocky Mountain Research Station.

Amorim, H. C. S., Ashworth, A. J., Moore, P. A., Wienhold, B. J., Savin, M. C., Owens, P. R., Jagadamma, S., Carvalho, T. S., & Xu, S. T. (2020a). Soil quality indices following long-term conservation pasture management practices. *Agriculture Ecosystems & Environment*, 301. doi:10.1016/j.agee.2020.107060

Amorim, H. C. S., Ashworth, A. J., Wienhold, B. J., Savin, M. C., Allen, F. L., Saxton, A. M., Owens, P. R., & Curi, N. (2020b). Soil quality indices based on long-term conservation cropping systems management. *Agrosystems, Geosciences & Environment*, 3, e20036. doi:10.1002/agg2.20036

Andrews, S. S., Flora, C. B., Mitchell, J. P., & Karlen, D. L. (2003). Growers' perceptions and acceptance of soil quality indices. *Geoderma*, 114, 187–213. doi:10.1016/S0016-7061(03)00041-7

Andrews, S. S., Karlen, D. L., & Cambardella, C. A. (2004). The soil management assessment framework: A quantitative soil quality evaluation method. *Soil Science Society of America Journal*, 68, 1945–1962. doi:10.2136/sssaj2004.1945

Apesteguía, M., Virto, I., Orcaray, L., Bescansa, P., Enrique, A., Imaz, M. J., & Karlen, D. L. (2017). Tillage effects on soil quality after three years of irrigation in Northern Spain. *Sustainability*, 9. doi:10.3390/su9081476

Arshad, M. A., & Coen, G. M. (1992). Characterization of soil quality: Physical and chemical criteria. *American Journal of Alternative Agriculture*, 7, 25–31. doi:10.1017/S0889189300004410

Cherubin, M. R., Karlen, D. L., Franco, A. L. C., Cerri, C. E. P., Tormena, C. A., & Cerri, C. C. (2016). A soil management assessment framework (SMAF) evaluation of Brazilian sugarcane expansion on soil quality. *Soil Science Society of America Journal*, 80, 215–226. doi:10.2136/sssaj2015.09.0328

Cherubin, M. R., Tormena, C. A., & Karlen, D. L. (2017). Soil quality evaluation using soil management assessment framework (SMAF) in Brazilian Oxisols with contrasting texture. *Rev. Bras. Cienc. Solo*, 41, e0160148. doi:10.1590/18069657rbcs20160148

Da Luz, F. B., Da Silva, V. R., Kochem Mallmann, F. J., Bonini Pires, C. A., Debiasi, H., Franchini, J. C., & Cherubin, M. R. (2019). Monitoring soil quality changes in diversified agricultural cropping systems by the soil management assessment framework (SMAF) in southern Brazil. *Agriculture, Ecosystems & Environment* 281, 100–110. doi:10.1016/j.agee.2019.05.006

Erkossa, T., Itanna, F., & Stahr, K. (2007). Indexing soil quality: A new paradigm in soil science research. *Australian Journal of Soil Research*, 45:129–137. doi:10.1071/SR06064

Frost, P. S. D., van Es, H. M., Rossiter, D. G., Hobbs, P., & Pingali, P. L. (2019). Soil health characterization of agricultural catchments in Jharkhand, India. *Applied Soil Ecology*, 138, 171–180. doi:10.1016/j.apsoil.2019.02.003

Gura, I., & Mnkeni, P. N. S. (2019a). Crop rotation and residue management effects under no till on the soil quality of a Haplic Cambisol in Alice, Eastern Cape, South Africa. *Geoderma*, 337, 927–934. doi:10.1016/j.geoderma.2018.10.042

Gura, I., & Mnkeni, P. N. S. (2019b). Dataset on the use of the soil management assessment framework (SMAF) in evaluating the impact of conservation agriculture strategies on soil quality. *Data in Brief*, 22, 578–582. doi:10.1016/j.dib.2018.12.044

Granatstein, D. (1990). Crop and soil management. In C. E. Beus (Ed.), *Prospects for sustainable agriculture in the Palouse: Farmer experiences and viewpoints* (pp. 5–34). XB 1016. Pullman, WA: Washington State University.

Hammac, W. A., Stott, D. E., Karlen, D. L., & Cambardella, C. A. (2016). Crop, tillage, and landscape effects on near-surface soil quality indices in Indiana. *Soil Science Society of America Journal*, 80, 1638–1652. doi:10.2136/sssaj2016.09.0282

Harris, R. F., & Bezdicek, D. F. (1994). Descriptive aspects of soil quality/health. In J. W. Doran & T. B. Parkin (Eds.), *Defining soil quality for a sustainable environment* (pp. 23–35). SSSA Special Publ. No. 35. Madison, WI: Soil Science Society of America.

Herrick, J. E., Lessard, V. C., Spaeth, K. E., Shaver, P. L., Dayton, R. S., Pyke, D. A., Jolley, L., & Goebel, J. J. (2010). National ecosystem assessments supported by scientific and local knowledge. *Frontiers in Ecology and the Environment*, 8, 403–408. doi:10.1890/100017

Herrick, J. E., Neff, J., Quandt, J., Salley, S., Maynard, J., Ganguli, A., & Bestelmeyer, B. (2019). Prioritizing land for investments based on short- and long-term land potential and degradation risk: A strategic approach. *Environmental Science Policy* 96, 52–58. doi:10.1016/j.envsci.2019.03.001

Herrick, J. E., Urama, K. C., Karl, J. W., Boos, J., Johnson, M. V. V., Shepherd, K. D., Hempel, J., Bestelmeyer, B. T., Davies, J., Guerra, J. L., Kosnik, C., Kimiti, D. W., Ekai, A. L., Muller, K., Norfleet, L., Ozor, N., Reinsch, T., Sarukhan, J., & West, L. T. (2013). The global land-potential knowledge system (LandPKS): Supporting evidence-based, site-specific land use and management through cloud computing, mobile applications, and crowdsourcing. *Journal of Soil Water Conservation*, 68, 5A–12A. doi:10.2489/jswc.68.1.5A

Idowu, O. J., van Es, H. M., Abawi, G. S., Wolfe, D. W., Ball, J. I., Gugino, B. K., Moebius, B. N., Schindelbeck, R. R., & Bilgili, A.V. (2008). Farmer-oriented assessment of soil quality using field, laboratory, and VNIR spectroscopy methods. *Plant Soil*, 307, 243–253. doi:10.1007/s11104-007-9521-0

Idowu, O. J., van Es, H. M., Abawi, G. S., Wolfe, D. W., Schindelbeck, R. R., Moebius-Clune, B. N., & Gugino, B. K. (2009). Use of an integrative soil health test for evaluation of soil management impacts. *Renewable Agricultural Food Systems*, 24, 214–224. doi:10.1017/S1742170509990068

Iqbal, M., van Es, H. M., Schindelbeck, R. R., & Moebius-Clune, B. N. (2012). Soil health indicators measure multiple benefits of farmyard manure application. *International Journal of Agricultural Biology*, 2012, 242–250.

Karlen, D. L., & Stott, D. E. (1994). A framework for evaluation physical and chemical indicators of soil quality. In J. W. Doran (Ed.), *Defining soil quality for a sustainable environment* (pp. 53–72). SSSA Special Publ. No. 35. Madison, WI: Soil Science Society of America.

Karlen, D. L., Stott, D. E., Cambardella, C. A., Kremer, R. J., King, K. W., & McCarty, G. W. (2014). Surface soil quality in five Midwestern cropland Conservation Effects Assessment Project watersheds. *Journal of Soil Water Conservation*, 69, 393–401. doi:10.2489/jswc.69.5.393

Karlen, D. L., Tomer, M. D., Neppel, J., & Cambardella, C. A. (2008). A preliminary watershed scale soil quality assessment in north central Iowa, USA. *Soil Tillage Research*, 99, 291–299. doi:10.1016/j.still.2008.03.002

Karlen, D. L., Veum, K. S., Sudduth, K. A., Obrycki, J. F., & Nunes, M. R. (2019). Soil health assessment: Past accomplishments, current activities, and future opportunities. *Soil Tillage Research*, 195, 104365. doi:10.1016/j.still.2019.104365

Liebig, M. A., Miller, M. E., Varvel, G. E., Doran, J. W., & Hanson, J. D. (2004). AEPAT: Software for assessing agronomic and environmental performance of management practices in long-term agroecosystem experiments. *Agronomy Journal*, 96, 109–115. doi:10.2134/agronj2004.0109

Lisboa, I. P., Cherubin, M. R., Satiro, L. S., Siqueira-Neto, M., Lima, R. P., Gmach, M. R., Wienhold, B. J., Schmer, M. R., Jin, V. L., Cerri, C. C., & Cerri, C. E. P. (2019). Applying soil management assessment framework (SMAF) on short-term sugarcane straw removal in Brazil. *Industrial Crops & Products*, 129, 175–184. doi:10.1016/j.indcrop.2018.12.004

McCord, S. E., Webb, N. P., Bonefont, K., Burke, R., Edwards, B., & Bestelmeyer, B. (2020). A data commons approach can help span scales and stakeholders to support ecosystem conservation and land use. Presentation at the Ecological Society of America Annual Meeting. 3–6 Aug. 2020.

Moebius-Clune, B. N., Moebius-Clune, D. J., Gugino, B. K., Idowu, O. J., Schindelbeck, R. R., Ristow, A. J., van Es, H. M., Thies, J. E., Shayler, H. A., McBride, M. B., Wolfe, D. W., & Abawi, G. S. (2016). *Comprehensive assessment of soil health*. The Cornell Framework Manual, edition 3.0. Ithaca, NY: Cornell University. Available at http://soilhealth.cals.cornell.edu/training-manual/.

Moebius-Clune, B. N., van Es, H. M., Idowu, O. J., Schindelbeck, R. R., Kimetu, J. M., Ngoze, S., Lehmann, J., & Kinyangi, J. M. (2011). Long-term soil quality degradation along a cultivation chronosequence in western Kenya. *Agriculture, Ecosystems & Environment*, 141, 86–99. doi:10.1016/j.agee.2011.02.018

Nunes, M. R., Karlen, D. L., & Moorman, T. B. (2020a). Tillage intensity effects on soil structure indicators—A US meta-analysis. *Sustainability*, 12, 1–17. doi:10.3390/su12052071

Nunes, M. R., Karlen, D. L., Moorman, T. B., & Cambardella, C. A. (2020b). How does tillage intensity affect chemical soil health indicators? A United States meta-analysis. *Agrosystems, Geosciences and Environment*, 3, e20083. doi:10.1002/agg2.20083

Nunes, M. R., Karlen, D. L., Veum, K. L., Moorman, T. B., & Cambardella, C. A. (2020c). Biological soil health indicators respond to tillage intensity: A US meta-analysis. *Geoderma*, 369, 114335. doi:10.1016/j.geoderma.2020.114335

Nunes, M. R., van Es, H. M., Schindelbeck, R. R., Ristow, A., & Ryan, M. (2018). No-till and cropping system diversity improve soil health and crop yield. *Geoderma*, 328, 30–43. doi:10.1016/j.geoderma.2018.04.031

Nunes, M. R., Veum, K. S., Parker, P. A., Holan, S. H., Karlen, D. L., Amsili, J. P., van Es, H. M., Wills, S. A., Seybold, C. A., & Moorman, T. B. (2021). The soil health assessment protocol and evaluation applied to soil organic C. *Soil Science Society of America Journal* (First look). https://doi.org/10.1002/saj2.20244

Pellant, M., Shaver, P. L., Pyke, D. A., Herrick, J. E., Lepak, N., Riegel, G., Kachergis, E., Newingham, B.A., . . ., Busby, F. E. (2020). Interpreting indicators of rangeland health, Version 5. Tech Ref 1734-6. Denver, CO: U.S. Department of the Interior, Bureau of Land Management, National Operations Center.

Pyke, D. A., Herrick, J. E., Shaver, P., & Pellant, M. (2002). Rangeland health attributes and indicators for qualitative assessment. *Journal of Range Management*, 55, 584–597. doi:10.2307/4004002

Quandt, A., Herrick, J. E., Peacock, G., Salley, S., Buni, A., Mkalawa, C. C., & Neff, J. (2020). A standardized land capability classification system for land evaluation using mobile phone technology. *Journal of Soil Water Conservation*, 75(5), 579–589. doi:10.2489/jswc.2020.00023

Rekik, F., van Es, H. M., Hernandez-Aguilera, J. N., & Gomez, M. (2018). Soil health assessment for coffee farms on andosols in Colombia. *Geoderma Regional*, 14, e01176. doi:10.1016/j.geodrs.2018.e00176

Romig, D. E., Garlynd, M. J., Harris, R. F., & McSweeney, K. (1995). How farmers assess soil health and quality. *Journal of Soil Water Conservation*, 50, 229–236.

Ruiz, F., Cherubin, M. R., & Ferreira, T. O. (2020). Soil quality assessment of constructed technosols: Towards the validation of a promising strategy for land reclamation, waste management and the recovery of soil functions. *Journal of Environmental Management*, 276, 111344. doi:10.1016/j.jenvman.2020.111344

Stott, D. E. (2019). *Recommended soil health indicators and associated laboratory procedures. Soil Health Technical Note No. 450-03*. Washington, DC: USDA, Natural Resources Conservation Service.

Stott, D. E., Andrews, S. S., Liebig, M. A., Wienhold, B. J., & Karlen, D. L. (2010). Evaluation of β-glucosidase activity as a soil quality indicator for the soil management assessment framework (SMAF). *Soil Science Society of America Journal*, 74, 107–119. doi:10.2136/sssaj2009.0029

Stott, D. E., Cambardella, C. A., Wolf, R., Tomer, M. D., & Karlen, D. L. (2011). A soil quality assessment within the Iowa River South Fork Watershed. *Soil Science Society of America Journal*, 75, 2271–2282. doi:10.2136/sssaj2010.0440

Stott, D. E., Karlen, D. L., Cambardella, C. A., & Harmel, R. D. (2013). A soil quality and metabolic activity assessment after fifty-seven years of agricultural management. *Soil Science Society of America Journal*, 77, 903–913. doi:10.2136/sssaj2012.0355

Swanepoel, P. A., Du Preez, C. C., Botha, P. R., Snyman, H. A., & Habig, J. (2015). Assessment of tillage effects on soil quality of pastures in South Africa with indexing methods. *Soil Research*, 53, 274–285. doi:10.1071/SR14234

USDA–NRCS. (2014). *National conservation practice standards*. Washington, DC: USDA, Natural Resources Conservation Service. Retrieved from www.nrcs.usda.gov/wps/portal/nrcs/detailfull/national/technical/cp/ncps/?cid=nrcs143_026849 (Accessed 10 Aug. 2020).

van Es, H. M., & Karlen, D. L. (2019). Reanalysis validates soil health indicator sensitivity and correlation with long-term crop yields. *Soil Science Society of America Journal*, 83, 721–732. doi:10.2136/sssaj2018.09.0338

Veum, K. S., Kremer, R. J., Sudduth, K. A., Kitchen, N. R., Lerch, R. N., Baffaut, C., Stott, D. E., Karlen, D. L., & Sadler, E. J. (2015). Conservation effects on soil quality indicators in the Missouri Salt River Basin. *Journal of Soil Water Conservation*, 70, 232–246. doi:10.2489/jswc.70.4.232

Wander, M. M., Walter, G. L., Nissen, T. M., Bollero, G. A., Andrews, S. S., & Cavanaugh-Grant, D. A. (2002). Soil quality: Science and process. *Agronomy Journal*, 94, 23–32. doi:10.2134/agronj2002.0023

Wienhold, B. J., Karlen, D. L., Andrews, S. S., & Stott, D. E. (2009). Protocol for soil management assessment framework (SMAF) soil indicator scoring curve development. *Renewable Agricultural Food Systems*, 24, 260–266. doi:10.1017/S1742170509990093

Wienhold, B. J., Pikul, J. L., Liebig, M. A., Mikha, M. M., Varvel, G. E., Doran, J. W., & Andrews, S. S. (2006). Cropping system effects on soil quality in the Great Plains: Synthesis from a regional project. *Renewable Agricultural Food Systems* 21, 49–59. doi:10.1079/RAF2005125

Wymore, A. W. (1993). *Model-based systems engineering: An introduction to the mathematical theory of discrete systems and to the tricotyledon theory of system design*. Boca Raton, FL:CRC Press.

Zobeck, T. M., Crownover, J., Dollar, M., Van Pelt, R. S., Acosta-Martínez, V., Bronson, K. F., & Upchurch, D. R. (2007). Investigation of soil conditioning index values for Southern High Plains agroecosystems. *Journal of Soil Water Conservation*, 62, 433–442.

Zobeck, T. M., Halvorson, A. D., Wienhold, B., Acosta-Martinez, V., & Karlen, D. L. (2008). Comparison of two soil quality indexes to evaluate cropping systems in northern Colorado. *Journal of Soil Water Conservation*, 63, 329–338. doi:10.2489/jswc.63.5.329

Zobeck, T. M., Steiner, J. L., Stott, D. E., Duke, S. E., Starks, P. J., Moriasi, D. N., & Karlen, D. L. (2015). Soil quality index comparisons using Fort Cobb, Oklahoma, watershed-scale land management data. *Soil Science Society of America Journal*, 79, 224–238. doi:10.2136/sssaj2014.06.0257

# 6

# Soil Health Assessment of Forest Soils

*Deborah S. Page-Dumroese, Felipe G. Sanchez, Ranjith P. Udawatta, Charles (Hobie) Perry, and Grizelle González*

## Introduction

Forest sustainability is explicitly tied to soil health, which has been defined as "the capacity of a soil to function within ecosystem boundaries to sustain biological productivity, maintain environmental quality, and promote plant and animal health" (Doran et al., 1996; Sigua, 2018). This definition includes the ability of soil to function effectively as a component of healthy forests (Schoenholtz et al., 2000) and is linked to the soils ability to support physical, chemical, and biological properties while also suppressing plant pathogens (van Bruggen and Semenov, 2000). In broad terms, forest soil health can be defined as a capacity for water retention, carbon (C) sequestration, and plant productivity, or it could simply be defined as the ability of the soil to produce biomass (Schoenholtz et al., 2000). For forested ecosystems to be sustainable, soil health must be maintained. Forest soil health is linked with the amount and composition of surface and mineral soil organic matter (SOM; Harvey et al., 1979; Harvey et al., 1981). In fact, the U.S. Department of Agriculture (USDA) Forest Service requirement to leave 25 to 27 tons $ha^{-1}$ of coarse woody material greater than 14 cm in diameter. comes from the need to provide 'parent material' for decayed wood in many forest ecosystems (Harvey et al., 1981) and ensure a healthy population of ectomycorrhizal fungi. Although mineral SOM is a small fraction of mineral soil mass (1–5%), it is responsible for a majority of soil physical, chemical, and biological properties through plant litter and anthropogenic inputs (Liang et al., 1998; Six and Jastrow, 2002). Since SOM

*Soil Health Series: Volume 1 Approaches to Soil Health Analysis*, First Edition.
Edited by Douglas L. Karlen, Diane E. Stott, and Maysoon M. Mikha.

also improves soil health, it also increases the chances for successful restoration after disturbance (Hagen-Thorn et al., 2004).

Forest and agroforest soils provide many ecosystem services including timber, clean water, flood control, and biodiversity, but maintaining soil health is difficult because of numerous stressors (*i.e.*, climate change, air pollution, altered water tables, intensive harvesting and site preparation, wildfire, invasive species, and overgrazing). No single forest soil health indicator is adequate because changes in one property will likely influence others. Therefore, using a variety of chemical, physical, and biological indicators (properties), land managers can better understand the impacts of stand- and watershed-scale manipulations, temperature and moisture variability, deep soil processes, and invasive species on soil health.

Evaluating forest soil health is difficult because soils are dynamic systems influenced by physical, chemical, and biological properties that are quantifiable using several appraisal techniques, many already being used to assess soil health. For example, the USDA Forest Service Forest Inventory and Analysis (FIA) program collects soil data during its inventory of the Nation's forest resources. Furthermore, many national forests use the Forest Soil Disturbance Monitoring Protocol (Page-Dumroese et al., 2009) to collect short- and long-term data on changes in soil physical attributes after land management, but routine measurements of multiple soil health indicators can be expensive. Therefore, remote sensing (Chaudhary et al., 2012) is often combined with in-field sensors to substitute for more expensive laboratory testing of physical, chemical, and biological properties (*e.g.*, Hemmat and Adamchuk, 2008; Sudduth et al., 2013). Recently, the Comprehensive Assessment of Soil Health (CASH) approach has been used in the eastern U.S. to measure 15 physical, biological, and chemical indicators using a scoring system (Fine et al., 2017). These efforts, and many others, are providing the baseline data needed to test and assess both soil- and ecosystem-health. Currently there is no universally accepted protocol for assessing soil health, but Table 6.1 lists several key soil chemical, physical, and biological properties that are widely used, with some being static (point-in-time) measures and others dynamic (process level) measures.

There are many different indicators that can be used to assess soil health, but those that are simple, easy to measure, relatively rapid to use, cover the largest number of soil types, and sensitive to environmental changes and land management are the most desirable (Doran and Zeiss, 2000; Knoepp et al., 2000). Herein, we discuss how soil health is being assessed in complex agroforest, tropical and temperate ecosystems. Additionally, we present a national perspective using FIA protocols.

**Table 6.1** Examples of soil physical, chemical, and biological properties that are used to assess temperate, agroforest, and tropical forest soil health.

| Indicator* | Reference | Comment |
|---|---|---|
| ***Soil Physical Properties*** | | |
| Visual assessment of surface soil changes | Page-Dumroese et al., 2009 | Rapid forest soil disturbance monitoring protocol |
| Soil compaction | Shestak and Busse, 2005; | Soil compaction linked to biological processes |
| Aggregate stability | Herrick et al., 2001 | Rapid field assessment kit |
| Porosity | Schoenholtz et al., 2000; Udawatta et al., 2006 | Including texture, aeration, runoff, infiltration, water holding capacity |
| Coarse fragments | Page-Dumroese et al., 1999; Jurgensen et al., 2017 | Importance of coarse-fragments for calculating nutrient pools and supporting logging equipment. |
| Water holding capacity | Schoenholtz et al., 2000 | Determines water flux, erosion, runoff, infiltration, storage |
| ***Soil Chemical Properties*** | | |
| Active C | Page-Dumroese et al., 2015 | Rapid field test |
| Organic C | Harris et al., 1996 | Specific scoring functions for plant productivity |
| | Busse et al., 2006; | Changes in fungal and bacterial biomass |
| | Sanchez et al., 2006a | |
| Organic matter | Gregorich et al., 1994; Laik et al., 2009; Wang and Wang, 2007 | Soil organic matter pools respond to changes in plant productivity, climate, and land use |
| **Nutrients** | | |
| Nitrogen (organic and mineral) | Doran and Parkin, 1994 | A primary indicator of soil health |
| Base cations (e.g., Calcium, magnesium, potassium) and Cation Exchange Capacity | Merilä et al., 2010 | With linkages to plants and soil microbial communities |
| Integrated physical and chemical measures | Amacher et al., 2007 | Integrates 19 measured physical and chemical properties into a single 'vital sign' of overall soil quality. |

**Table 6.1** (Continued)

| Indicator* | Reference | Comment |
|---|---|---|
| Salinity (electrical conductivity) | Doran and Parkin, 1994 | Basic indicator of soil health |
| ***Soil biological Properties*** | | |
| Decomposition of standard substrates | Jurgensen et al., 2006; González et al., 2008 | Index of organic matter decay as influenced by biotic and abiotic factors |
| Fauna | van Straalen, 1998; Knoepp et al., 2000; González and Seastedt, 2001 | Bioindicator of soil health |
| PFLA, DNA or RNA-based techniques | van Bruggen and Semenov, 2000 | Microbial diversity and function, species richness, disease suppression |
| Microbial techniques combined with organic matter and nutrient analyses | Arias et al., 2005 | |

*Linkages to forest soil health are too numerous to list, only a select few are noted here.

# Ecosystem Examples

## Agroforestry

Agroforestry (AF) is an intensive land management practice where trees and shrubs are integrated into crop and/or livestock management practices to optimize numerous benefits arising from biophysical interactions among the components (Gold and Garrett, 2009). Five main AF practices are: riparian buffers, alley cropping, windbreaks, silvopasture, and forest farming. Riparian buffers exist around water bodies while upland buffers are mostly located on contours to create alley cropping. Windbreaks protect crops, livestock, and farm structures from wind and snow. Silvopasture is the integration of trees, forage, and livestock and is designed to produce a high-value timber product, while providing short-term cash flow from livestock (Klopfenstein et al., 1997). Furthermore, AF practices were approved by both the afforestation and reforestation programs and under the Clean Development Mechanisms of the Kyoto Protocol for C sequestration (IPCC, 2007; Watson et al., 2000; Smith et al., 2007). However, current literature lacks information on the role of AF practices on soil health. This section will highlight benefits of AF practices on soil health parameters including soil C, physical, biological, and chemical soil properties and a soil's capacity to degrade harmful chemicals and promote biodiversity.

## Carbon Sequestration

A decrease of soil C causes degradation of soil health that could lead to a food insecurity and declining ecosystem sustainability (Godfray et al., 2010; Montgomery, 2010). Agroforestry practices increase soil C and reduce greenhouse gases (Schoeneberger et al., 2012a; Udawatta and Jose, 2012; Stefano and Jacobson, 2018) because perennial vegetation stores more C in above- and belowground biomass, soil, living and dead organisms, and root exudates (Cairns and Meganck, 1994; Pinho et al., 2012) as compared to row crops or grazing. Since both forest and grassland C sequestration and storage patterns are active in AF ecosystems, a higher percentage of C is allocated to belowground biomass through an extended growing season (Schroeder, 1993; Kort and Turnock, 1999; Sharrow and Ismail, 2004; Morgan et al., 2010). Diverse vegetation also promotes diverse soil communities (fauna and flora), development of surface and deep roots, and reduced soil disturbance which, combined, enhance C sequestration potential (Udawatta et al., 2009; Kumar et al., 2010; Paudel et al., 2011; Udawatta and Jose, 2012). In addition, SOM concentrations are greater at the soil surface (0–15 cm) and near the base of trees as compared with soil located greater distances from perennial vegetation or deeper in the soil profile (Seiter et al., 1995; Sauer et al., 2007; Fig. 6.1). Brandle et al. (1992) estimated that 22.2 metric tons of

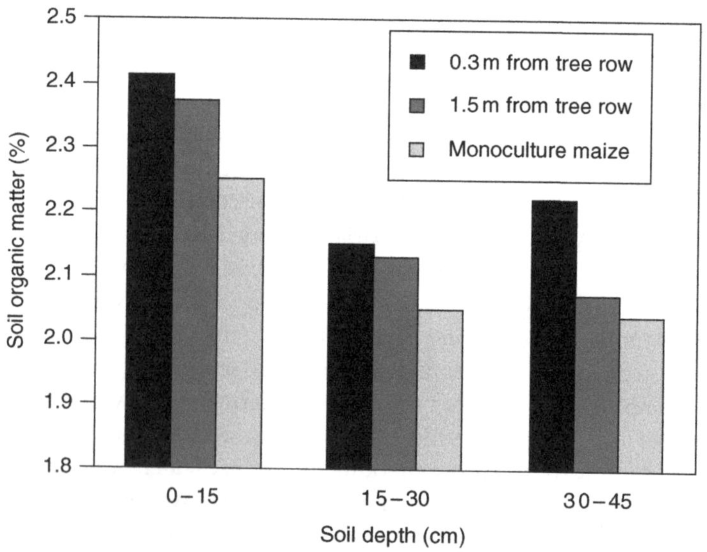


**Figure 6.1** Soil organic matter percentage decreased with increasing distance and depth from tree rows for a 4-year old red alder-corn alley cropping system in western Oregon, USA. (Adapted from Seiter et al., 1995).

C is stored on 1.96 million ha of shelterbelts and is a model for enhanced sequestration to mitigate climate change.

In the United States, pasture and grazing lands occupy 266 and 52 million ha, respectively with the potential to sequester as much as 516 Tg C $yr^{-1}$ just by converting 10% of the pasture lands to silvopasture and 10% of the crop land to alley cropping (Nair et al., 2009). Furthermore, Udawatta and Jose (2012) have estimated that silvopasture, alley cropping, windbreaks, and riparian buffers could sequester 642 Tg C $yr^{-1}$ in the United States (Fig. 6.2).

## Soil Physical Indicators

Climate change is expected to increase the intensity of rainfall in the 21st century increasing soil erosion 16 and 58% (Nearing et al., 2004). This predicted climatic shift emphasizes the importance of soil conservation. By using AF practices, soil health can be increased through improved soil bulk density, aggregate stability, porosity, water holding capacity, infiltration, and limiting sediment movement (Seobi et al., 2005; Udawatta et al., 2006; 2009; 2011a; Adhikari et al., 2014). Aggregate stability is greater in AF soils as compared to soil under row crops or in grazed lands (Udawatta et al., 2008; Paudel et al., 2011, 2012) and can lead to a more stable SOM pool (Novara et al., 2012). Bulk density in AF sites was reduced by 2.3% after six years with a concomitant increase in porosity (Seobi et al., 2005). These changes in soil bulk density, porosity, and SOM also serve to increase infiltration, saturated hydraulic conductivity, water holding capacity, and water storage (Kumar, 2012; Akdemir et al., 2016; Alagele et al., 2018) resulting in enhanced production of food, fiber and, thus, soil health (Balandier et al., 2008; Dimitriou et al., 2009; Udawatta et al., 2011a).

## Soil Biological Indicators

Soil fauna composition, microbial activity, microbial biomass, and enzyme activity are good soil health indicators that can be used to predict land management effects on water, microbes, nutrient use efficiency, and disease suppression (Bandick and Dick, 1999; Boerner et al., 2000; Schloter et al., 2003; Brussaard et al., 2007). Soil enzymes are greater in AF soils as compared to row crop and grazed lands (Mungai et al., 2006; Udawatta et al., 2009; Paudel et al., 2011) because of improved litter quality and quantity, diverse vegetation, and root exudates. In addition, a diverse microbial community can sequester eight to ten times more C than monoculture systems (Polgase et al., 2008). These changes imply positive effects on soil biochemical processes and microbial resilience which ultimately leads to greater soil health, resilience, and productivity (Rivest et al., 2013).

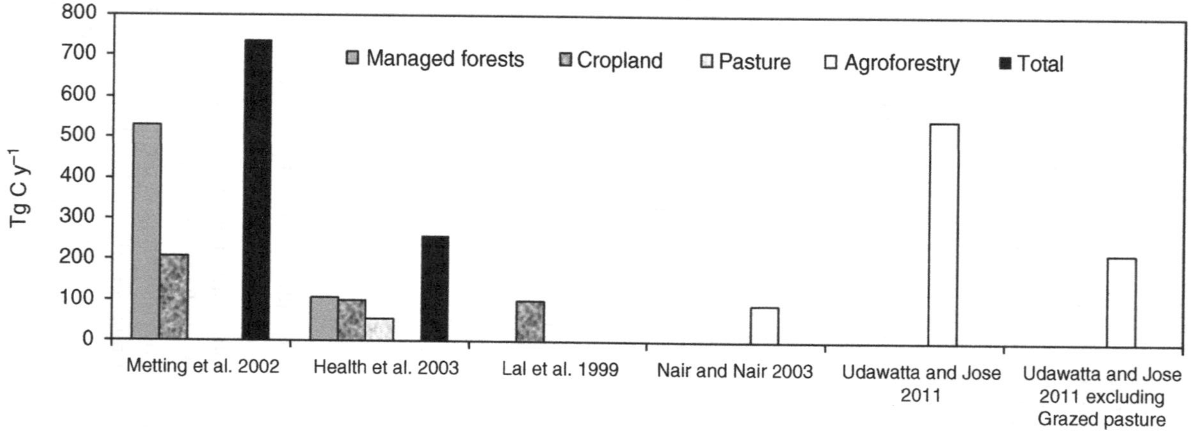


**Figure 6.2** Carbon sequestration potential for various management systems in the USA (*Source:* Udawatta and Jose, 2012).

## Soil Enrichment and Decontamination

Nutrient additions, long-term productivity, sustainability, and the reduction of water pollution and hypoxia conditions all enrich soil functions (Jose, 2009; Udawatta et al., 2009; Zomer et al., 2009; Udawatta et al., 2011b). Soil enrichment occurs through filtering of nutrients and sediment within the root zone and in the reduction of water erosion and sediment losses (Udawatta et al., 2011b; Allen et al., 2004). Agroforestry practices retain nutrients and C by filtering nutrients and sediment and reducing water erosion and these properties increase as buffer width increases (Broadmeadow and Nisbet, 2004; Schultz et al., 2009; Udawatta et al., 2011b).

There are many sources of soil contamination (*e.g.*, mining, industrialization, rapid urbanization, herbicides, pesticides, antibiotics, personal care products) and phytoremediation is a cost effective, noninvasive, and socially preferred approach to remove environmental contaminants (Boyajian and Carreira, 1997). Fast growing tree species such as poplars (*Populous* spp) and forage grasses (e.g., *Panicum vigatum)* can produce large amounts of biomass and deep roots that can both tolerate and extract large amounts of contaminants through plant uptake (Dhillon et al., 2008; Gomes, 2012; Zalesny et al., 2019). To date, more than 400 plant species have been identified that can accumulate heavy metals. These plants remove a contaminant from the soil and accumulate the contaminants in shoots and/or roots. This helps reduce contamination in the soil and increase soil health (Paz-Ferreiro et al., 2014).

Key soil health benefits associated with changes in soil properties by AF operations are reduced water pollution, enhanced soil microbial population and diversity, and increased C sequestration which also result in healthier ecosystems and land productivity. Indicators such as bulk density, porosity, infiltration rate, and microbial diversity can help track changes with the AF ecosystems and show reduced water loss and erosion, climate change mitigation, and enhanced ecosystem resilience.

## Tropical Forests

As with AF systems, tropical soil health reflects the interaction of physical, chemical, and biological components, but the relative importance of those properties differs depending on local climate and vegetation. There are two types of tropical forests: Moist/wet (2000 to > 8000 mm of precipitation $yr^{-1}$) and dry (several months of severe drought). Tropical forests occur about 25° north and south of the Equator and have both evergreen and deciduous tree species. In tropical forests rainfall seasonality, distribution, and variability drive soil moisture, litter accumulation and decay, soil respiration, and overall productivity. Further, the length of the wet season will, in part, dictate the amount of SOM storage (Rohr et al., 2013). Threats to soil health in tropical systems include a changing climate, fire, hurricanes, and land conversion (Jaramillo and Murray-Tortarolo, 2019; Cusack and Marín-Spiotta, 2019).

To understand the drivers of tropical soil health, it is critical to understand the rates of decomposition and incorporation of organic material to determine the capacity of an ecosystem to sequester C and cycle nutrients important for productivity, fertility, and overall ecosystem health. In tropical ecosystems, climate may be less important than the biological regulation by soil macro-fauna (Lavelle et al., 1993; Heneghan et al., 1999; González and Seastedt, 2001). Soil macrofauna are more common in the tropics than in temperate zones while soil microfauna are more abundant in the temperate regions (González and Seastedt, 2000; González, 2002). This latitudinal variation in the types of micro- and macrofauna and their relative importance can have a significant effect on litter breakdown rates. Consequently, biological properties including the diversity of micro- and macrofauna are an important determinant of soil health in the tropics (González and Lodge, 2017). In addition, the abundance of various soil fauna also changes with latitude (Swift et al., 1979).

## Environmental Gradients and Future Climate Projections

Henareh Khalyani et al. (2016) assessed different general circulation models and greenhouse gas emission scenarios of downscaled climate projections to inform future climatology and its potential impacts to tropical regions in the U.S., namely the Caribbean islands. Those projections indicate a reduction in precipitation and an increase of 4 to 9°C in air temperature. In addition, they projected a high likelihood of shifts in ecological life zones to drier conditions. The combination of decreased rainfall, increasing variability of rainfall, and higher air temperatures would lead to reductions in soil moisture and changes to SOM dynamics. Though microbial soil processes will likely adjust to changes in rainfall, additional stressors of climate change may lower microorganism diversity or productivity, thus reducing microbial pool resiliency (Silver, 1998). Consequently, tropical forest soils may be affected by the changing climate through increased variability in SOM decay and potential changes to soil biota, oxygen concentrations, and nutrient accessibility (González et al., 2013).

Additional research at the Luquillo Experimental Forest in Puerto Rico suggests that threats of a changing climate to forest soil health vary along elevational gradients. In a field soil translocation experiment, Chen et al. (2017) studied the impacts of decreasing temperature but increasing moisture on soil organic C and respiration along an elevation gradient in northeastern Puerto Rico. Soils translocated from low- to high- elevation showed an increased respiration rate with decreased soil organic C content, which suggested that the increased soil moisture and altered soil microbes may affect respiration rates. Further, soils translocated from high- to low-elevation also showed an increased respiration rate with reduced soil organic C, suggesting that the higher temperature at low elevations enhanced decomposition rates. Thus, tropical soils at high elevations may be at risk of releasing sequestered C into the atmosphere giving a warming climate in the Caribbean (Chen et al., 2017).

In tropical forests, seasonal soil decomposition is closely tied to wet and dry cycles, suggesting that seasonal adjustments in temperature and moisture due to climate change are likely to affect decomposer communities, soil resource quantity and distribution, and litter quality (Silver, 1998). Decomposer organisms can be key determinants of decay in Puerto Rico (e.g., González and Seastedt, 2001). Yet, the contribution of different groups of decomposers to the decay of coarse woody debris, might vary among the different forest types located along elevation and environmental gradients (González and Luce, 2013). For example, González and Luce (2013) found the decay of coarse woody debris was most strongly correlated with white rot fungi in cloud forests (tropical wet forests) located at the tops of mountains (high elevation). In contrast, wood decay rates in tropical dry forests (low elevation) was related to the high diversity of species and functional groups of wood-inhabiting animals (Torres and González, 2005, González et al., 2008). Thus, the distribution of particular groups of organisms might be more important predictors of wood decay in tropical regions than climatic constraints (González, 2002, 2016).

## Tropical Soil Chemical and Biological Properties

Tropical forests are places where large quantities of debris are periodically generated during tropical storms and hurricanes. Such disturbances may increase nutrient losses from the forest depending on how the debris is managed, how the microbiota responds to the disturbance, and the chemical and physical characteristics of the soil (Miller and Lodge, 1997). Canopy disturbances associated with severe hurricane storms dramatically alter the physicochemical environment and the amounts of debris deposited into the forest floor (Lodge et al., 1991; Ostertag et al., 2003; Shiels and González, 2014). In addition, canopy disturbances alter the patterns in litterfall and associated nutrient cycling (Scatena and Lugo, 1995; Lugo and Scatena, 1996); hurricane litter contains a high proportion of green leaves from which nutrients have not been translocated, thus altering the litter quality in the forest floor (Richardson et al., 2010). Cascading effects from canopy openness can account for most of the shifts in the forest biota and biotic processes, which include increased plant recruitment and richness, as well as the decreased abundance and diversity of several animal groups (Richardson et al., 2010; Shiels et al., 2015). Opening the canopy decreases litterfall and litter moisture, thereby inhibiting lignin-degrading fungi, decreasing litter invertebrate richness, diversity, and biomass, and ultimately slowing decomposition (González et al., 2014; Lodge et al., 2014; Shiels et al., 2015). Yet, modeling exercises relate the long-term effect of hurricane generated debris to a positive effect of decaying large woody debris on soil P exchange capacity (Sanford et al., 1991; Zimmerman et al., 1995).

Decaying wood may impact the physical, chemical, and biotic properties of the underlying soil (Zalamea et al., 2016), stabilize soil temperature (Spears et al., 2003), and contribute to the spatial heterogeneity of soil formation and resultant nutrient cycling in tropical forests (Zalamea et al., 2007, 2016). Further, tree species and decay stage are important factors defining the effect of decaying wood on the distribution of available nutrients (Zalamea et al., 2016). Lodge et al. (2016) found that surface soil on the upslope side of the logs can have significantly more nitrogen (N) and microbial biomass, likely from accumulation of leaf litter above the logs on steep slopes. To summarize, tropical cyclones deposit coarse woody debris on forest floors and significantly alter soil C and N dynamics, which consequently alter soil fertility, soil health, and forest productivity (Lodge et al., 2016).

## Earthworms as Bioindicators

The occurrence or abundance of soil fauna can be considered a soil health bioindicator as it can reflect some habitat characteristics. These non-anthropogenic disturbances may increase nutrient losses from tropical forests, depending on how the debris is managed, how soil organisms respond to disturbance, and the chemical and physical characteristics of soil and litter (González and Barberena-Arias, 2017). Earthworms are recognized as indicators of soil fertility and health because they play an active role in organic matter movement and decay, soil formation, and improvement of soil structure by channeling and bioturbation (Fragoso and Lavelle, 1992; Liu and Zou, 2002). Their relatively large size (ranging from 1 to 80 cm, or larger), slow displacement in soil, and ability to re-colonize sites make earthworm concentrations and diversity easy to measure and an attractive bioindicator of soil health (Paoletti, 1999).

Tropical land-use changes affect the abundance and community structure of earthworms. Converting tropical forests to pastures often results in the reduction of aboveground plant litter inputs, causing the disappearance of soil surface litter layer (Zou and González, 1997; Paoletti, 1999). In short-term field experiments, manipulating plant litter inputs lead to a decrease in anecic worms (those that build permanent burrows in the mineral soil; González and Zou, 1999; Sánchez and Zou, 2004). Furthermore, deforestation and establishment of exotic grasses decreases the diversity of earthworm communities in tropical Oxisols and Ultisols (Zou and González, 1997; Sánchez et al., 2003). Native earthworm communities are often negatively affected by non-native tree species, but they can be preserved in plantations where native tree species are planted (Zou and González, 2001). Conventional practices of site preparation and harvesting favors nonnative soil dwelling earthworms which often have a deleterious effect on native litter-dwelling worms. Therefore, forest management practices can drastically alter earthworm populations and diversity, and yet, maintaining a healthy population of earthworms can further promote forest nutrition and soil health in tropical tree plantations (Zou and González, 2001).

## Temperate Forests

Temperate forests, located at mid-latitudes north and south of the Equator, are comprised of both evergreen and deciduous tree species and influenced by strong seasonal temperature shifts and other climate differences. Tree species, climate, parent material, and topography all influence temperate forest soil formation (Binkley and Fisher, 2012), but overstory species often influence soil chemistry (e.g., pH), biology (litter decomposition rate and rooting depth), and soil available water (Adams et al., 2019).

A number of natural and anthropogenic threats make temperate forest soil health vulnerable to degradation. One of the greatest concerns is environmental change due to catastrophic fires, but since temperate forests are often found near population centers, soil health can also be threatened by N deposition, acid rain, and invasive earthworms, plants, insects, and diseases. Forest management affects soil C storage through harvesting and site preparation operations that significantly alter surface and subsurface physical, chemical, and biological properties. However, in temperate and other ecosystems, if the external stress is not too great and the frequency and severity of disturbance are low, many soil properties will return to pre-disturbance conditions if given enough time (Morris et al., 1997).

Similar to AF and tropical forest soils, an important indicator of temperate forest soil health is SOM. This was documented by a 1958 Calhoun Experimental Forest study and gradually became a way to restore forest cover to land previously damaged by agriculture throughout the southeastern U.S (Metz, 1958). The long-term dataset documented the effect trees had on building surface organic horizons and improving soil moisture retention (Richter and Markewitz, 2001). Further, increasing C inputs led to even higher rates of decomposition (Richter et al., 1999), and also soil changed porosity and nutrient cycling, thus generally improving soil health.

## The North American Long-Term Soil Productivity Study

One important program for forest soil health is the Long-Term Soil Productivity (LTSP) study (Mushinski et al., 2017; Powers et al., 2005). This coordinated network of over 100 sites (Fig. 6.3) was initiated to address concerns that SOM removal and compaction were causing declines in temperate forest soil health. In general, loss of branches and twigs from the site did not alter tree growth, but when the surface organic horizons were removed many site experienced declines in productivity. Further, the effects of harvesting, compaction, and SOM removal varied considerably from site-to-site.

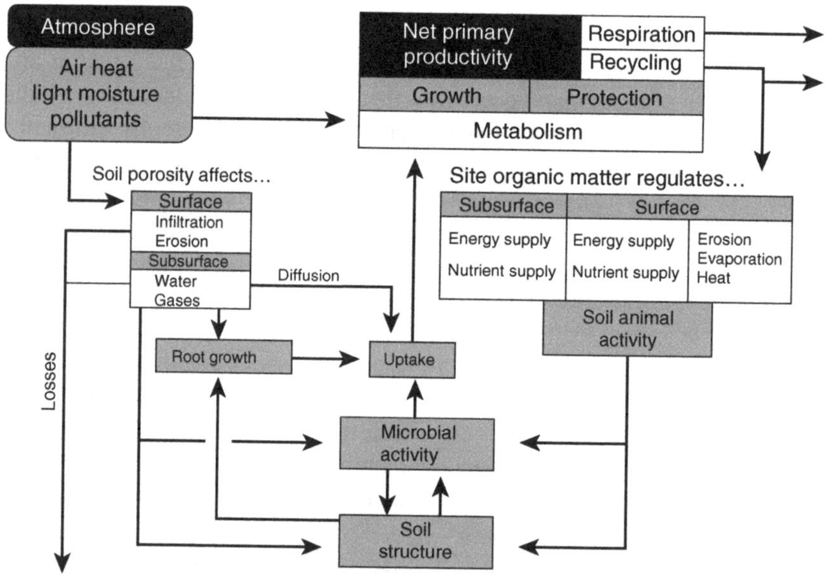


**Figure 6.3** Geographic extent of LTSP sites in North America.

## Temperate Forest Soil Health

Because temperate forests are widely distributed, a decline in soil health is likely to be of global importance. A rise in temperature of 1–2°C will have regional impacts on precipitation amounts and patterns leading to a changes in soil temperature and moisture properties (Adams et al., 2019). These changes, coupled with land use change, air pollution, and biotic effects will control forest productivity, SOM decomposition, and the C balance within the soil and in the atmosphere (IPCC, 2003).

### Elevated Carbon Dioxide

Rising carbon dioxide ($CO_2$) is considered to be a major driver of climate change and can significantly affect forest growth, SOM, and soil health. For example, the Free-Air Carbon Dioxide ($CO_2$) Experiments (FACE) study sites have shown that an increase in $CO_2$ can increased forest productivity, but there was no evidence to suggest that C storage increased in mineral soils beneath temperate forests (Norby et al., 2002). This is likely due to increased soil respiration (Phillips et al., 2012), root turnover (Bader et al., 2009), and microbial activity (Larson et al., 2002). It has also been shown that litter quality and species changes can change the quality of C inputs to the soil (MacKenzie et al., 2004). Changes in atmospheric and soil C associated with changing climate emphasizes the need to maintain soil bulk density, aeration, surface organic horizons, and other properties that promote soil aggregation and stable nutrient cycling.

### Fire

In many temperate forests, wildfire is the most severe threat to soil health. With persistent and recurring drought, often coupled with high temperatures, wildfire risk and severity have been increasing and have resulted in greater loss of all or part of the surface organic horizons and mineral SOM. Those effects cascade into sediment loss, loss of C storage, and degradation of aggregate stability, but they may be partially off-set by creation of pyrogenic C. In fire-prone ecosystems, DeLuca and Aplet (2008) estimate that pyrogenic C inputs may account for 15 to 20% of the total C in temperate, coniferous forest mineral soils, but subsequent harvesting or thinning activities may reduce this amount. A recent meta-analysis noted an overall increase in C in frequently burned forests, but this varies by ecosystem type and burn severity (Pellegrini et al., 2017).

### Thinning or Bioenergy Harvests

Many temperate forest stands need restoration because of lack of harvesting, fire suppression, and insect or disease outbreaks have resulted in excess woody biomass within many stands. There is also recent interest in using forests for bioenergy feedstock which may increase harvest operations on many sites. Little is known regarding the impact on soil health of repeated harvest due to forest thinning operations or feedstock extraction, but loss of SOM from periodic stand disturbances can be either negligible (Sanchez et al., 2006b) or significant, depending on soil type, tree species, ecosystem, or climatic regime (Grigal and Vance, 2000).

Conversely, excess biomass left during thinning or bioenergy harvest may provide fuel for uncharacteristically severe wildfires (Page-Dumroese et al., 2010).

Temperate forests supply important ecosystem services and therefore, it is critical to maintain a healthy, productive soil. Many nations have strong forest inventory and monitoring programs that also incorporates soil data collection. These inventories provide an opportunity to be pro-active in response to stressors that may alter forest or soil health.

## Using National Forest Inventory and Analysis Data to Assess Forest Soil Health

In 1928, the McSweeney- McNary Forest Research Act (P.L. 70–466) directed the U.S. Department of Agriculture Forest Service to make "... a comprehensive survey of the present and prospective requirements for timber and other forest products of the United States..." The first inventories were completed in the 1930s and focused on the economic value of the forest by documenting the extent and status of timber resources (Cowlin and Moravets, 1938; Cunningham and Moser, 1938; Spillers, 1939). Seventy years later the Agriculture Research, Extension, and Education Reform Act of 1998 (16 USC 1642(e)) mandated that the Forest Service Forest Inventory and Analysis (FIA) program "make available to the public a report, prepared in cooperation with State foresters, that ... contains an analysis of forest health conditions and trends." This Act resulted in the development of comprehensive sampling protocols designed to monitor forest soils (chemical and physical properties), down and dead wood, lichens, ozone damage, tree crown condition, and vegetation diversity (O'Neill et al., 2005a; O'Neill et al., 2005b; Woodall et al., 2011).

Soil sampling conducted by the FIA program differs from the USDA National Cooperative Soil Survey (NCSS) in several critical ways. The FIA program is based on providing a spatially balanced, statistical sample of the landscape (Reams et al., 2005). In contrast, NCSS identifies relatively homogenous map units for the purpose of sampling (Soil Science Division Staff, 2017). Although digital soil mapping provided by the NCSS facilitates the estimation of error or uncertainty associated with soil properties (Kienast-Brown et al., 2017), the design-based framework used by FIA allows calculation of statistically robust estimates of various attributes along with associated estimates of uncertainty (Scott et al., 2005). Additionally, because of the explicit focus on the forest resource, FIA has a much greater sampling intensity across the forested landscape, under both public and private ownerships (Table 6.2). In contrast to the NCSS use of soil scientists who describe genetic horizons and sample the soil using soil pits (Schoeneberger et al., 2012b), FIA field crews collect ocular estimates of soil properties (erosion and rutting) and sample soils adjacent to the field plot by depth (0–10 and

**Table 6.2** Forest soil sampling intensity of FIA plots by forest-type group.

| Forest-type group | Number of plots |
|---|---|
| Alder/maple | 27 |
| Aspen/birch | 447 |
| California mixed conifer | 65 |
| Douglas-fir | 387 |
| Elm/ash/cottonwood | 365 |
| Exotic hardwoods | 16 |
| Exotic softwoods | 13 |
| Fir/spruce/mountain hemlock | 351 |
| Hemlock/Sitka spruce | 69 |
| Loblolly/shortleaf pine | 356 |
| Lodgepole pine | 164 |
| Longleaf/slash pine | 75 |
| Maple/beech/birch | 718 |
| Nonstocked | 229 |
| Oak/gum/cypress | 142 |
| Oak/hickory | 1573 |
| Oak/pine | 253 |
| Other eastern softwoods | 33 |
| Other hardwoods | 46 |
| Other softwoods | 1 |
| Other western softwoods | 76 |
| Pinyon/juniper | 814 |
| Ponderosa pine | 253 |
| Redwood | 2 |
| Spruce/fir | 307 |
| Tanoak/laurel | 18 |
| Tropical hardwoods | 135 |
| Western larch | 15 |
| Western oak | 85 |
| Western white pine | 2 |
| White/red/jack pine | 163 |
| Woodland hardwoods | 328 |
| **Grand Total** | **7528** |

10–20 cm) by using a slide hammer and volumetric soil core sampler whenever possible (USDA Forest Service, 2011). Both organizations submit their samples to laboratories for physical and chemical analyses. FIA samples soil in association with a comprehensive sample of the aboveground forest resource (USDA Forest Service, 2017) to facilitate our understanding of linkages between soil and forest health (O'Neill et al., 2005b).

Observed soil properties are extrapolated by NCSS using map units. FIA does not define homogenous units for the purposes of sampling or extrapolation. Instead, it relies on two statistical strategies for estimation. The first method uses the base sample and the underlying sample in a design-based framework to convert point observations to estimates (Scott et al., 2005). The second method implements statistical imputation techniques to convert point observations to continuous surfaces (Wilson et al., 2012; Wilson et al., 2013; Domke et al., 2016; Domke et al., 2017).

Soils data collected by the FIA program have been used in a number of different assessments, either in isolation or in combination with other attributes, ranging from regional to national scales.

For example, the Forest Service is responsible for producing the official forest C estimates submitted to the UN Framework Convention on Climate Change (US Environmental Protection Agency, 2018). While soil C stocks have been reported since the early 1990s, they were initially estimated without the benefit of field observations on the FIA plot network. Estimates were based on linkages between FIA plots and NCSS map units (Smith and Heath, 2002, Amichev and Galbraith, 2004). With the addition of the soil indicator to the FIA program in 1999, the foundation was laid for reporting forest C stocks by using continuous, integrated field monitoring (Perry et al., 2009). Forest floor and mineral soil C stocks are currently estimated using an imputation approach (Domke et al., 2016, Domke et al., 2017). In a testament to the value of the inventory, Domke et al. (2016) demonstrated that current Good Practice Guidance for Tier 1 approaches (estimates based on simples methods and default values) overestimate forest floor C stocks. These empirical data have also demonstrated the importance of reforestation for C sequestration (Nave et al., 2018).

In addition to the FIA program, the Forest Service has a Forest Health Monitoring program that plays a role nurturing thoughtful investigations of forest health, including soil data. Their annual National Technical Reports serve as venues to explore nascent trends detected across the monitoring network. Early reports summarized evaluations of soil C and other physical and chemical properties (Perry and Amacher, 2007a; Perry and Amacher, 2007b; Perry and Amacher, 2007c). Building on these assessments of individual soil properties,

Amacher et al. (2007) developed a technique to integrate the multiple chemical and physical observations from FIA plots into an index of forest soil health. Furthermore, FIA data has been used to map the legacy of atmospheric deposition observed in Ca:Al ratios (Perry and Amacher, 2012) and increased mortality of sugar maple (*Acer saccharum;* Perry and Zimmerman, 2012). These myriad analyses illustrate how FIA has become a foundation for national forest resource assessment (Perry and Amacher, 2009).

While there are tremendous strengths in the Forest Service's monitoring of forest soils, it is important to acknowledge known limitations. First, the soil program is considered 'Core Optional', and is not implemented across the nation on a regular basis (Fig. 6.4). This limits the program's ability to map soil health properties of interest (e.g., SOC, N) and document change. Inference from this inventory program is limited by the sampling protocol. Fixed depth sampling is a reproducible method of data collection, but it may yield samples straddling soil horizons and mixing soils of divergent properties (Schoeneberger et al., 2012b); this complicates interpretation of the resulting data and estimates. Finally, sampling frequency could be optimized to detect changes of interest to the Forest Service and partners concerned about managing forest health. The intensity of FIA's field campaigns currently yield a complete sample of the forest resource every 5 to 7 yr in the eastern U. S. and every 10 yr in the western U.S. Because the annual FIA program was implemented in stages beginning in the late 1990s, sampling is not necessarily completed uniformly across years. Sampling is also paused between inventory cycles to increase the likelihood of capturing changes in soil properties.

Despite these limitations, soil sampling conducted by the FIA program represents a tremendously valuable, statistically sound sample of forest soil health. How might it be improved? The most common concerns fit broadly under sampling intensity. First, the mineral soil is sampled to 20 cm by a bulk density sampler where possible. However, IPCC Good Practice Guidance suggests monitoring of soil carbon to at least 30 cm (IPCC, 2003) in agricultural soils. However, deeper sampling (at least to 80 cm) in forested soils should be conducted to better assess deep soil C pools and changes over time (Harrison et al., 2011). Second, only one sample of mineral soil is collected on each plot. This efficient use of limited funds provides landscape-level information, but it provides no detail on small-scale variation in soil health. Third, soil samples are collected only on a subset of the full plot network; originally soil sampling represented a 1/16th subset, but the program is exploring the value of sampling at greater intensities. Another major concern is the narrow focus on physical and chemical properties when microbes are now understood to have a critical role in forest diversity and productivity (van der Heijden et al., 2008). Sampling

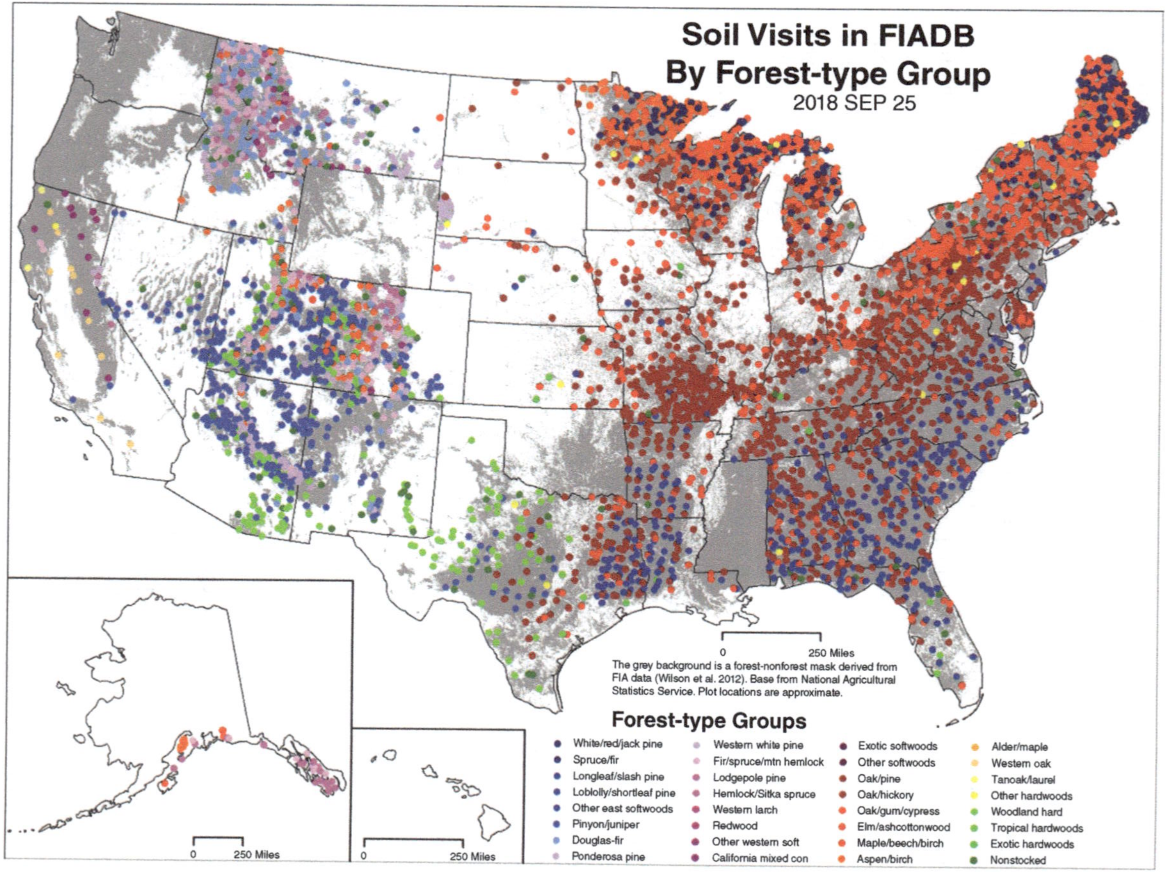


**Figure 6.4** Forest Inventory and Analysis samples collected by Forest-type Group.

soil metagenomics to understand fungal diversity may be a relatively inexpensive way to understand this impactful biological property (Tringe et al., 2005, Fierer et al., 2012) and has been piloted on FIA plots in northern Idaho (Ross-Davis et al., 2016).

## Forest Soil Health Data Limitations and Management Implications

Often, forest health measurements look at only aboveground responses because they are easier to measure than belowground responses. Numerous studies indicate that forest management and inherent soil factors will elicit differing tree responses (*e.g.,* Greacen and Sands, 1980; Senyk and Craigdallie, 1997; Heninger et al., 2001; Slesak et al., 2017). Ideally, above- and belowground data are needed at a site before harvest operations are conducted so that the magnitude of change and the functions and processes affected can be quantified (Grigal and Vance, 2000). Further, many studies constrain their sampling efforts to the surface mineral soil and to the fine fraction (< 2mm) and omit coarse wood, large rocks, or roots from sampling because of financial or time limitations. Recent studies have pointed out that coarse-fragments (Jurgensen et al., 2017) and deep soil nutrient pools and OM should also be considered to evaluate long-term impacts on soil health (Harrison et al., 2011). Not accounting for these factors could result in faulty soil health assessments (Slesak et al., 2017). In many cases baseline information might not be available. In this case, local specialists, use of the Natural Resource Conservation Service databases, or use of information from similar sites elsewhere may be necessary to make inferences on soil productivity changes. In general, low fertility, coarse-textured soils are at greater risk of nutrient limitations from land management than higher fertility soils with finer textures (Garrison et al., 2000).

Long-term studies are key to being able to link belowground ecological changes associated with land management. Metagenomics, standard decomposition substrates, soil fauna, and microbial biomass are all techniques that help link soil physical, chemical, and biological responses. Data gathered immediately after harvesting are a valuable tool, and there are many examples of developing risk rating systems for forest sites (e.g., Reeves et al., 2012). In fact, best management practices (BMPs) have been developed by some states that minimize or avoid soil impacts considered detrimental to forest productivity. Risk rating tools can provide a framework that, with local calibration, can be used across a wide variety of forested landscapes to depict soils that may be at risk of damage during ground-based harvest activities (Reeves et al., 2012). Use of a standardized soil monitoring protocol is also useful for assessing short- and long-term soil health based on several soil quality indices (Heninger et al.,

2001; Page-Dumroese et al., 2010). This data will be useful for meta-analyses that examine soil and forest health changes.

### Climate Change, Fire Shifts, Invasive Species

Three points are clear: soil C is a pervasive material within all forested soils, it is crucial for providing ecosystem services (*e.g.*, soil water quality and quantity), and it is an essential indicator of soil health. Current US federal policy is to harvest forests in a manner that protects soil, watershed, fish, wildlife, recreations, and esthetic resources. Consequently, soil health must also be protected to ensure all other values are maintained. Since soil C is critical, we must begin to assess its vulnerability to climate, fire, and invasive species shifts and to understand these changes more widely.

Fire can affect soil C by changing the quantity and quality of C inputs by mineralizing surface OM and altering mineral soil C (Neary et al., 1999). However, those changes may be offset by creation of black C during wildfires, prescribed burns, or through the addition of biochar to forest soils (Page-Dumroese et al., 2018). Since many public lands have a short window for burning unmerchantable woody material, alternative markets such as bioenergy or bio-based products are one way to reduce the amount of residual woody material while simultaneously conserving C.

Linked to changes in soil C and N is an increase in invasive species. The initial increases in invasive species is caused by a chronic disruption in N, SOM, or nutrient cycling (Hobbs and Huenneke, 1992). Working to adjust these soil imbalances may be one method for restoring microbial-fauna-soil-plant relationships and foster increased soil health.

## Conclusion: Criteria and Indicators for Monitoring Forest Soil Health

Monitoring forest soil health is a process to estimate changes in soil conditions that have occurred since the last time it was measured. However, this approach gives no indication of future soil conditions that may result from continuing impacts of degrading processes (*e.g.*, climate change, pollution; Wagenet and Hutson, 1997). Our roles as forest soil scientists should be to anticipate effects in a prospective manner rather than retrospectively (Wagenet and Hutson, 1997; Adams et al., 2000). We must stretch our knowledge of soil data to encompass dynamic processes that underpin soil health assessments. Data from each of the previous examples can be used to give a qualitative perspective on the impact of management scenarios on soil health and provides resources that further our understanding. Information and data from both short- and long-term studies can be placed into decision trees that help integrate soil property changes into site-specific land management decisions (Wagenet and Hutson, 1997). Current measurements can also be combined with archived samples from numerous sources to help provide additional historical context about how soils are impacted over time by a changing climate and/or land management activities.

Researchers take static measurements of forest soil properties (cation exchange capacity, C, base cations, etc.), but it is imperative to also determine the cause and effect relationships between management and soil properties. These static measurements can also be used to develop risk rating systems. Risk rating systems used to develop BMPs are one way to use data to describe acceptable management retrospectively. Once these relationships are understood we can identify indicators of soil change that could lead to a decline in soil health and forest growth. From the empirical trials we can then move toward forecasting acceptable management across a wide-range of soil types and required ecosystem services.

Use of a standardized soil monitoring protocol is also essential for assessing soil health based on several soil characteristics (compaction, rutting, displacement, erosion, ground cover, burn severity; Heninger et al., 2001; Page-Dumroese et al., 2009). This data can then be used with meta-analyses that examine soil and forest health changes.

## Summary

- Sustainable management of temperate and tropical forests as well as AF sites depends on healthy soils and the ability to identify soil change indicators that reflect soil health declines.
- There are no widely-applicable standardized measurements or methods for assessing forest soil health.
- Soil texture influences how compaction or SOM loss can alter soil health.
- Faunal and microbial inventories, and the development of specialized taxonomic expertise, is needed to better describe organisms and biological property changes and links with aboveground changes.
- FIA forest and soil data can be used as an index of ecosystem health.
- Soil monitoring of management practices will help elucidate if we are meeting criteria for sustainability.
- Using long-term trials and archived samples we can begin to forecast ecosystem processes changes that may require a change in management.

## Acknowledgments

GG was supported by National Science Foundation to the Luquillo Critical Zone Observatory (EAR-1331841) and the Luquillo Long Term Ecological Research Program (DEB 1239764). The USDA Forest Service International Institute of Tropical Forestry, and The University of Puerto Rico provided additional support. Any trade, product, or firm name is used for descriptive purposes only and does not imply endorsement by the U.S. Government.

## References

Adams, M.B., Burger, J.A., Jenkins, A.B., and Zelazny, L. (2000). Impact of harvesting and atmospheric pollution on nutrient depletion of eastern US hardwood forests. *For. Ecol. Manage.* 138, 301–319. doi:10.1016/S0378-1127(00)00421-7

Adams, M.B., Kelly, C., Kabrick, J., and Schuler, J. (2019). Temperate forest and soils. In: Busse, M., Giardino, C., Morris, D., and Page-Dumroese, D., editors, *Global change and forest soils: Cultivating stewardship of a finite natural resource* (p. 83–108). Developments in Forest Soils Vol. 36. London: Elsevier. doi:10.1016/B978-0-444-63998-1.00006-9

Adhikari, P., Udawatta, R.P., Anderson, S.H., and Gantzer, C.J. 2014. Soil thermal properties under prairies, conservation buffers, and corn/soybean land use systems. *Soil Sci. Soc. Am. J.* 78, 1977–1986. doi:10.2136/sssaj2014.02.0074

Akdemir, E., Anderson, S.H., and Udawatta, R.P. (2016). Influence of agroforestry buffers on soil hydraulic properties relative to row crop management. *Soil Sci.* 181, 368–376. doi:10.1097/SS.0000000000000170

Alagele, S.M., S.H. Anderson, R.P. Udawatta, and S. Jose. 2018. Agroforestry, grass, biomass crop, and row-crop management effects on soil water dynamics for claypan landscapes. p. 212–216 In N. Ferreiro-Dominguez and M.R. Mosquera-Losada (Eds.) 4th European Agroforestry Conference, Agroforestry as Sustainable Land Use. 28-30 May 2018. Nijmegen, The Netherlands.

Allen, S.C., Jose, S., Nair, P.K.R., Brecke, B.J., Nkedi-Kizza, P., and Ramsey, C.L. (2004). Safety-net role of tree roots: Evidence from a pecan (Carya illinoensis K. Koch)–cotton (Gossypium hirsutum L.) alley cropping system in the southern United States. *For. Ecol. Manage.* 192, 395–407. doi:10.1016/j.foreco.2004.02.009

Amacher, M.C., O'Neill, K.P., and Perry, C.H. (2007). *Soil vital signs: A new soil quality index (SQI) for assessing forest soil health.* RMRS-GTR-65. Fort Collins, CO: US Department of Agriculture, Forest Service, Rocky Mountain Research Station. doi:10.2737/RMRS-RP-65

Amichev, B.Y., and Galbraith, J.M. (2004). A revised methodology for estimation of forest soil carbon from spatial soils and forest inventory data sets. *Environ. Manage.* 33(Suppl. 1), S74–S86. doi:10.1007/s00267-003-9119-0

Arias, M.E., González-Pérez, J.A., González-Vila, F.J., and Ball, A.S. (2005). Soil health-a new challenge for microbiologist and chemists. *Int. Microbiol.* 8, 13–21.

Bader, M., Hiltbrunner, E., and Körner, C. (2009). Fine root responses of mature deciduous forest trees to free air carbon dioxide enrichment (FACE). *Funct. Ecol.* 23, 913–921. doi:10.1111/j.1365-2435.2009.01574.x

Balandier, P., de Montard, F., and Curt, T. (2008). Root competition for water between trees and grass in a silvopastoral plots of 10 year old *Prunus avium*. In D.R. Batish, S. Jose, H. Pal Singh, R.K. Kohli, (eds.), *Ecological basis of agroforestry* (p. 253–270). New York: CRC Press.

Bandick, A.K., and Dick, R.P. (1999). Field management effects on soil enzyme activities. *Soil Biol. Biochem.* 31, 1471–1479. doi:10.1016/S0038-0717(99)00051-6

Binkley, D., and Fisher, R. (2012). *Ecology and management of forest soils*. New York: John Wiley & Sons.

Boerner, R.E.J., Decker, K.L.M., and Sutherland, E.K. (2000). Prescribed burning effects on soil enzyme activity in a southern Ohio hardwood forest: A landscape-scale analysis. *Soil Biol. Biochem.* 32, 899–908. doi:10.1016/S0038-0717(99)00208-4

Brandle, J.R., Wardle, T.D., and Bratton, G.F. 1992. Opportunities to increase tree planting in shelterbelts and the potential impacts on carbon storage and conservation (p. 157–176). In N.R. Sampson and D. Hair (eds.), *Forests and Global Change Vol.1 Opportunities for increasing forest cover*. Washington, D.C.: American Forests.

Broadmeadow, S., and Nisbet, T.R. (2004). The effects of riparian forest management on the freshwater environment: A literature review of best management practice. *Hydrol. Earth Syst. Sci.* 8, 286–305. doi:10.5194/hess-8-286-2004

Boyajian, G.E., and Carreira, L.H. (1997). Phytoremediation: A clean transition from laboratory to marketplace? *Nat. Biotechnol.* 15, 127–128. doi:10.1038/nbt0297-127

Brussaard, L., deRuiter, P.C., and Brown, G.G. (2007). Soil biodiversity for agricultural sustain-ability. *Agric. Ecosyst. Environ.* 121, 233–244. doi:10.1016/j.agee.2006.12.013

Busse, M.D., Beattie, S.E., Powers, R.F., Sanchez, F.G., and Tiarks, A.E. (2006). Microbial community responses in forest mineral soil to compaction, organic matter removal, and vegetation control. *Can. J. For. Res.* 36, 577–588. doi:10.1139/x05-294

Cairns, M.A., and Meganck, R.A. (1994). Carbon sequestration, biological diversity, and sustainable development: Integrated forest management. *Environ. Manage.* 18, 13–22. doi:10.1007/BF02393746

Chen, D., Yu, M., González, G., Zou, X., and Gao, Q. 2017. Climate impacts on soil carbon processes along an elevation gradient in the tropical Luquillo Experimental Forest. *Forests* 8(3), 90. doi:10.3390/f8030090

Chaudhary, W.P., Sudduth, K.A., Kitchen, N.R., and Kremer, R.J. (2012). Reflectance spectroscopy detects management and landscape differences in soil carbon and nitrogen. *Soil Sci. Soc. Am. J.* 76, 597–606. doi:10.2136/sssaj2011.0112

Cowlin, R.W., and Moravets, F.L. (1938). *Forest statistics for eastern Oregon and eastern Washington from inventory phase of forest survey*. PNW Old Series Research Notes No. 25. Portland, OR: U.S. Department of Agriculture, Forest Service, Pacific Northwest Research Station.

Cunningham, R.N., and Moser, H.C. (1938). *Forest areas and timber volumes in the Lake States*. Econ. Notes No. 10. St. Paul, MN: U.S. Department of Agriculture, Forest Service, Lake States Forest Experiment Station.

Cusack, D., and Marín-Spiotta, E. (2019). Wet tropical soils and global change. In M. Busse, C. Giardino, D. Morris, and D. Page-Dumroese, (eds.), *Global change and forest soils: Cultivating stewardship of a finite natural resource* (p. 131–169). Amsterdam, The Netherlands: Elsevier. doi:10.1016/B978-0-444-63998-1.00008-2

DeLuca, T.H., and Aplet, G.H. (2008). Charcoal and carbon storage in forest soils of the Rocky Mountain West. *Front. Ecol. Environ* 6, 18–24. doi:10.1890/070070

Dhillon, K.S., Dhillon, S.K., and Thind, H.S. (2008). Evaluation of different agroforestry tree species for their suitability in the phytoremediation of seleniferous soils. *Soil Use Manage.* 24, 208–216. doi:10.1111/j.1475-2743.2008.00143.x

Dimitriou, I., Busch, G., Jaconbs, S., P. Schmidt-Walter, and N. Lambersdorf. 2009. A review of the impact of short rotation coppice cultivation on water issues. *Landbauforschung Volkenrode* 59:197–206.

Domke, G.M., Perry, C.H., Walters, B.F., Woodall, C.W., Russell, M.B., and Smith, J.E. (2016). Estimating litter carbon stocks on forest land in the United States. *Sci. Total Environ.* 557-558, 469–478. doi:10.1016/j.scitotenv.2016.03.090

Domke, G.M., Perry, C.H., Walters, B.F., Nave, L.E., Woodall, C.W., and Swanston, C.W. (2017). Toward inventory-based estimates of soil organic carbon in forests of the United States. *Ecol. Appl.* 27(4), 1223–1235. doi:10.1002/eap.1516

Doran, J.W., and Parkin, T.B. (1994). Defining and assessing soil quality. In: J.W. Doran, (ed.), *Defining soil quality for a sustainable environment, Proceedings of a symposium*. Madison, WI: SSSA Special Publication Number 35. p. 1–21. doi:10.2136/sssaspecpub35.c1

Doran, J.W., Sarrantonio, M., and Liebig, M.A. (1996). Soil health and sustainability. *Adv. Agron.* 56, 2–54.

Doran, J.W., and Zeiss, M.R. (2000). Soil health and sustainability: Managing the biotic component of soil quality. *Appl. Soil Ecol.* 15, 3–11. doi:10.1016/S0929-1393(00)00067-6

Fierer, N., J.W. Leff, B.J. Adams, U.N. Nielsen, S.T. Bates, C.L. Lauber, S. Owens, J.A. Gilbert, D.H. Wall, and J.G. Caporoso. (2012). Cross-biome metagenomic analyses of soil microbial communities and their functional attributes. In *Proceedings of the National Academy of Sciences of the United States of America* 109(52), 21390–21395. doi:10.1073/pnas.1215210110

Fine, A.K., van Es, H.M., and Schindelbeck, R.R. 2017. Statistics, scoring functions, and regional analysis of a comprehensive soil health database. *Soil Sci. Soc. Am. J.* 81, 589–601. doi:10.2136/sssaj2016.09.0286

Fragoso, C., and Lavelle, P. (1992). Earthworm communities of tropical rain forests. *Soil Biol. Biochem.* 24(12), 1397–1408. doi:10.1016/0038-0717(92)90124-G

Garrison, M.T., Moore, J.A., Shaw, T.M., and Mika, P.G. (2000). Foliar nutrient and tree growth response of mixed-conifer stands to three fertilization treatments in north east Oregon and north central Washington. *For. Ecol. Manage.* 132, 183–198. doi:10.1016/S0378-1127(99)00228-5

Gregorich, E.G., Monreal, C.M., Carter, M.R., Angers, D.A., and Ellert, B. (1994). Towards a minimum data set to assess soil organic matter quality in agricultural soils. *Can. J. Soil Sci.* 74, 367–385. doi:10.4141/cjss94-051

Grigal, D.F., and Vance, E.D. (2000). Influence of soil organic matter on forest productivity. *N. Z. J. For. Sci.* 30, 169–205.

Godfray, H.C.J., Beddington, J.R., Crute, I.R., Haddad, L., Lawrence, D., Muir, J.F., Pretty, J., Robinson, S., Thomas, S.M., and Toulmin, C. (2010). Food security: The challenge of feeding 9 billion people. *Science* 327(5967)812–818. doi:10.1126/science.1185383

Gold, M.A., and Garrett, H.E. (2009). Agroforestry nomenclature, concepts, and practices. In H.E. Garrett, (ed.), *North American agroforestry: An integrated science and practice* (p. 45–56). 2nd ed. New York: Wiley. doi:10.2134.2009.northamericanagroforestry.2ed.ch.

Gomes, H.I. (2012). Phytoremediation for bioenergy: Challenges and opportunities. *Environ. Tech. Rev.* 1, 59–66. doi:10.1080/09593330.2012.696715

González, G. (2002). Soil organisms and litter decomposition. In: R.S. Ambasht and N.K. Ambasht, editors, *Modern trends in applied terrestrial ecology* (p. 315–329). London, U.K.: Kluwer Academic/Plenum Publishers. doi:10.1007/978-1-4615-0223-4_16

González, G. (2016). Deadwood, soil biota and nutrient dynamics in tropical forests: A review of case studies from Puerto Rico. In C. McCown et al. (eds). *Proceedings of the One Hundred Twelfth Annual Meeting of the American Wood Protection Association* (p. 206–208). Vol. 112. 1-3 May 2016. San Juan, Puerto Rico. Birmingham, Alabama: American Wood Protection Association.

González, G., and Barberena-Arias, M.F. (2017). Ecology of soil arthropod fauna in tropical forests: A review of studies from Puerto Rico. *J. Agric. Univ. P R.* 101(2), 185–201.

González, G., and Lodge, D.J. (2017). Soil biology research across latitude, elevation and disturbance gradients: A review of forest studies from Puerto Rico during the past 25 years. *Forests* 8, 178–193. doi:10.3390/f8060178

González, G., and Luce, M.M. (2013). Woody debris characterization along an elevation gradient in.northeastern Puerto Rico. *Ecol. Bull.* 54, 181–193.

González, G., and Zou, X. (1999). Plant litter influences on earthworm abundance and community structure in a tropical wet forest. *Biotropica* 31(3), 486–493. doi:10.1111/j.1744-7429.1999.tb00391.x

González, G., and Seastedt, T.R. (2000). Comparison of the abundance and composition of litter fauna in tropical and subalpine forests. *Pedobiologia* 44, 545–555. doi:10.1078/S0031-4056(04)70070-0

González, G., and Seastedt, T.R. (2001). Soil fauna and plant litter decomposition in tropical and subalpine forests. *Ecol.* 82(4):955–964. doi:10.1890/0012-9658(2001)082[0955:SFAPLD]2.0.CO;2

González, G., Gould, W.A., Hudak, A.T., and Hollingsworth, T. (2008). Decay of aspen (*Populus tremuloides* Michx.) wood in moist and dry boreal, temperate, and tropical forest fragments. *Ambio* 37, 588–597. doi:10.1579/0044-7447-37.7.588

González, G., Waide, R.B., and Willig, M.R. (2013). Advanced in the understanding of spatiotemporal gradients in tropical landscapes: A Luquillo focus and global perspective. *Ecol. Bull.* 54, 245–250.

González, G., Lodge, D.J., Richardson, B.A., and Richardson, M.J. (2014). A canopy trimming experiment in Puerto Rico: The response of litter decomposition and nutrient release to canopy opening and debris deposition in a subtropical wet forest. *For. Ecol. Manage.* 332, 32–46. doi:10.1016/j.foreco.2014.06.024

Greacen, E.L., and Sands, R. (1980). Compaction of forest soils: A review. *Soil Res.* 18, 163–189. doi:10.1071/SR9800163

Hagen-Thorn, A., Callesen, I., Armolaitis, K., Monreal, C.M., and Ellert, B.H. (2004). The impact of six European tree species on the mineral topsoil in forest plantations on former agricultural land. *For. Ecol. Manage.* 195:373–384. doi:10.1016/j.foreco.2004.02.036

Harris, R.F., Karlen, D.L., and Mulla, D.J. (1996). A conceptual framework for assessment and management of soil quality and health. In: J.W. Doran and A.J. Jones, editors, *Methods for assessing soil quality* (p. 61–82). *Special Publication No. 49. Madison, WI: Soil Science Society of America.*

Harrison, R.B., Footen, P.W., and Strahm, B.D. (2011). Deep soil horizons: Contribution and importance to soil carbon pools and in assessing whole-ecosystem response to management and global change. *For. Sci.* 57(1), 67–76.

Harvey, A.E., Larsen, M.J., and Jurgensen, M.F. (1979). Comparative distribution of ectomycorrhizae in soils of three western Montana forest habitat types. *For. Sci.* 25, 350–358.

Harvey, A.E., Jurgensen, M.F., and Larsen, M.J. (1981). Organic reserves: Importance to ectomycorrhizae in forest soils of western Montana. *For. Sci.* 27, 442–445.

Heath, L.S., Kimble, J.M., Birdsey, R.A., and Lal, R. (2002). The potential of U.S. forest soils to sequester carbon. In: J.M. Heath, et al., editors, The potential of

US forest soils to sequester carbon and mitigate the greenhouse effects. Boca Raton, FL: CRC Press, p. 385–394. doi:10.1201/9781420032277-23

Hemmat, A., and Adamchuk, V.I. (2008). Sensor systems for measuring soil compaction: Review and analysis. *Comput. Electron. Agric.* 63, 89–103. doi:10.1016/j.compag.2008.03.001

Henareh Khalyani, A., Gould, W.A., Harmsen, E., Terando, A., Quiñones, M., and Collazo, J.A. (2016). Climate change implications for tropical islands: Interpolating and interpreting statistically Downscaled GCM Projections for Management and Planning. *J. Appl. Meteorol. Climatol.* 55(2), 265–282. doi:10.1175/JAMC-D-15-0182.1

Heneghan, L., Coleman, D.C., Zou, X., Crossley, D.A., and Haines, B.L. (1999). Soil microarthropod contributions to decomposition dynamics: Tropical-temperate comparisons of a single substrate. *Ecol.* 80, 1873–1882.

Heninger, R., Scott, W., Dobkowski, A., Miller, R., Anderson, H., and Duke, S. 2002. Soil disturbance and 10-year growth responses of coast Douglas-fir on non-tilled and tilled skid trails in the Oregon Cascades. *Can. J. For. Res.* 32, 233–246. doi:10.1139/x01-195

Herrick, J.E., Whitford, W.G., De Soyza, A.G., Van Zee, J.W., Havstad, K.M., Seybold, C.A., and Walton, M. (2001). Field soil aggregate stability kit for soil quality and rangeland health evaluations. *Catena* 44, 27–35. doi:10.1016/S0341-8162(00)00173-9

Hobbs, R.J., and Huenneke, L.F. (1992). Disturbance, diversity, and invasion: Implications for conservation. *Conserv. Biol.* 6, 324–337. doi:10.1046/j.1523-1739.1992.06030324.x

IPCC. (2003). Good practice guidance for land use, land-use change and forestry. Hayama, Kanagawa, Japan, IPCC National Greenhouse Gas Inventories Programme Technical Support Unit. Geneva, Switzerland: IPCC. https://www.ipcc-nggip.iges.or.jp/public/gpglulucf/gpglulucf_contents.html (Accessed 24 September 2018).

IPCC. (2007). Intergovernmental Panel on Climate Change 2007. Synthesis Report. Geneva, Switzerland: IPCC. http://www.ipcc.ch/pdf/assessment-report/ar4/syr/ar4_syr.pdf (Accessed 24 Sept. 2018).

Jaramillo, V.J., and Murray-Tortarolo, G.N. (2019). Tropical dry forest soils: Global change and local-scale consequences for soil biogeochemical processes. In M. Busse, C. Giardino, D. Morris, and D. Page-Dumroese, editors, *Global change and forest soils: Cultivating stewardship of a finite natural resource* (p. 109–130). Amsterdam, The Netherlands: Elsevier. doi:10.1016/B978-0-444-63998-1.00007-0

Jose, S. (2009). Agroforestry for ecosystem services and environmental benefits: An overview. *Agrofor. Syst.* 76, 1–10. doi:10.1007/s10457-009-9229-7

Jurgensen, M., D. Reed, D. Page-Dumroese, P. Laks, A. Collins, G. Mroz, and M. Degórski. (2006). Wood strength loss as a measure of decomposition in northern forest mineral soil. *Eur. J. Soil Biol.* 42(1), 23–31. doi:10.1016/j.ejsobi.2005.09.001

Jurgensen, M.F., D.S. Page-Dumroese, R.E. Brown, J.M. Tirocke, C.A. Miller, J.B. Pickens, and M. Wang. (2017). Estimating carbon and nitrogen pools in a forest soil: Influence of soil bulk density methods and rock content. *Soil Sci. Soc. Am. J.* 81, 1689–1696. doi:10.2136/sssaj2017.02.0069

Kienast-Brown, S., Z. Libohova, and J. Boettinger. 2017. Digital soil mapping. In C. Ditzler, K. Scheffe, and H.C. Monger (eds.), Soil survey manual (p. 295–354). USDA Handbook 18. Washington, D.C., Government Printing Office.

Klopfenstein, N.B., W.J. Rietveld, R.C. Carman, T.R. Clason, S.H. Sharrow, G. Garrett, and B.E. Anderson. (1997). *Silvopasture: An agroforestry practice*. Agroforestry Note 8. Fort Collins, CO: U.S. Department of Agriculture, Forest Service, Rocky Mountain Research Station.

Knoepp, J.D., Coleman, D.C., Crossley, Jr., D.A., and Clark, J.S. (2000). Biological indices of soil quality: An ecosystem case study of their use. *For. Ecol. Manage.* 138, 357–368. doi:10.1016/S0378-1127(00)00424-2

Kort, J., and Turnock, R. (1999). Carbon reservoir and biomass in Canadian Prairie Shelterbelts. *Agrofor. Syst.* 44, 175–186. doi:10.1023/A:1006226006785

Kumar, S., Udawatta, R.P., and Anderson, S.H. (2010). Root length density and carbon content of agroforestry and grass buffers under grazed pasture systems in a Hapludalf. *Agrofor. Syst.* 80, 85–96. doi:10.1007/s10457-010-9312-0

Kumar, C.P. (2012). Climate change and its impact on groundwater resources. *Int. J. Eng. Sci.* 1, 43–60.

Laik, R., Kumar, K., Das, D.K., and Chaturvedi, O.P. (2009). Labile soil organic matter pools in a calciorthent after 18 years of afforestation by different plantations. *Appl. Soil Ecol.* 42(2), 71–78. doi:10.1016/j.apsoil.2009.02.004

Larson, J.L., Zak, D.R., and Sinsabaugh, R.L. (2002). Extracellular enzyme activity beneath temperate trees growing under elevated carbon dioxide and ozone. *Soil Sci. Soc. Am. J.* 66, 1848–1856. doi:10.2136/sssaj2002.1848

Lavelle, P., Blanchart, E., and Martin, S. (1993). A hierarchical model for decomposition in terrestrial ecosystems: Application to soils of the humid tropics. *Biotropica* 25, 130–150. doi:10.2307/2389178

Liang, B.C., MacKenzie, A.F., Schnitzer, M., Monreat, C.M., Voroney, P.R., and Beyaert, R.P. (1998). Management-induced changes in labile soil organic matter under continuous corn in eastern Canadian soils. *Biol. Fertil. Soils* 26, 88–94. doi:10.1007/s003740050348

Liu, Z.G., and Zou, X. (2002). Exotic earthworms accelerate plant litter decomposition in a Puerto Rican pasture and wet forest. *Ecol. Appl.* 12, 1406–1417. doi:10.1890/1051-0761(2002)012[1406:EEAPLD]2.0.CO;2

Lodge, D.J., Scatena, F.N., Asbury, C.E., and Sánchez, M.J. (1991). Fine litterfall and related nutrient inputs resulting from Hurricane Hugo in subtropical wet and lower montane rain forests of Puerto Rico. *Biotropica* 23, 336–342. doi:10.2307/2388249

Lodge, D.J., Cantrell, S.A., and González, G. (2014). Effects of canopy opening and debris deposition on fungal connectivity, phosphorus movement between litter cohorts and mass loss. *For. Ecol. Manage.* 332, 11–21. doi:10.1016/j.foreco.2014.03.002

Lodge, D.J., Winter, D., González, G., and Cum, N. (2016). Effects of hurricane-felled tree trunks on soil carbon, nitrogen, microbial biomass, and root length in a wet tropical forest. *Forests* 7, 264. doi:10.3390/f7110264

Lugo, A.E., and Scatena, F.N. (1996). Background and catastrophic tree mortality in tropical moist, wet, and rain forests. *Biotropica* 28, 585–599. doi:10.2307/2389099

MacKenzie, M.D., DeLuca, T.H., and Sala, A. (2004). Forest structure and organic horizon analysis along a fire chronosequence in the low elevation forests of western Montana. *For. Ecol. Manage.* 203, 331–343. doi:10.1016/j.foreco.2004.08.003

Merilä, P., Malmivaara-Lämsä, M., Spetz, P., Stark, S., Vierikko, K., Derome, J., and Fritze, H. (2010). Soil organic matter quality as a link between microbial community structure and vegetation composition along a successional gradient in a boreal forest. *Appl. Soil Ecol.* 46, 259–267. doi:10.1016/j.apsoil.2010.08.003

Metz, L.J. (1958). Moisture held in pine litter. *J. For.* 56, 36.

Miller, R.M., and Lodge, D.J. (1997). Fungal responses to disturbance– agriculture and forestry. In K. Esser, Paul A. Lemke, editors, *The mycota, environmental and microbial relationships* (p. 65–84). Vol. V. Amsterdam, Springer Verlag.

Montgomery, H.L. (2010). *How is soil made?* New York: Crabtree Publishing.

Morgan, J.A., R.F. Follett, L.H. Allen, S.D. Grosso, J.D. Derner, F. Dijkstra, A. Franzluebbers, R. Fry, K. Paustian, and M.M. Schoeneberger. (2010). Carbon sequestration in agricultural land of the United States. *J. Soil Water Conserv.* 65, 6A–13A. doi:10.2489/jswc.65.1.6A

Morris, D.M., Kimmins, J.P.H., and Duckert, D.R. (1997). The use of soil organic matter as a criterion of the relative sustainability of forest management alternatives: A modeling approach using FORECAST. *For. Ecol. Manage.* 94, 61–78. doi:10.1016/S0378-1127(96)03984-9

Mungai, N.W., Motavalli, P.P., Kremer, R.J., and Nelson, K.A. (2006). Spatial variation of soil enzyme activities and microbial functional diversity in temperate alley cropping practices. *Biol. Fertil. Soils* 42, 129–136. doi:10.1007/s00374-005-0005-1

Mushinski, R.M., Boutton, T.W., and Scott, D.A. (2017). Decadal-scale changes in forest soil carbon and nitrogen storage are influenced by organic matter removal during timber harvest. *J. Geophys. Res. Biogeosci.* 122(4), 846–862. doi:10.1002/2016JG003738

Nair, P.K.R., Kumar, B.M., and Nair, V.D. (2009). Agroforestry as a strategy for carbon sequestration. *J. Plant Nutr. Soil Sci.* 172, 10–23. doi:10.1002/jpln.200800030

Nave, L.E., G.M. Domke, K.L. Hofmeister, and U. Mishra., [et al.]. 2018. Reforestation can sequester two petagrams of carbon in US topsoils in a century." *Proceedings of the National Academy of Sciences* 115(11): 2776–2781. doi:10.1073/pnas.1719685115

Nearing, M.A., Pruski, F.F., and O'Neal, M.R.. (2004). Expected climate change impact on soil erosion rates: A review. *J. Soil Water Conserv.* 59, 43–50.

Neary, D.G., Klopatek, C.C., DeBano, L.F., and Folliott, P.F. (1999). Fire effects on belowground sustainability: A review and synthesis. *For. Ecol. Manage.* 122, 51–71. doi:10.1016/S0378-1127(99)00032-8

Norby, R.J., P.J. Hanson, E.G. O'Neill, T.J. Tschaplinski, J.F. Weltzin, R.A. Hansen, W. Chang, S.D. Wullschleger, C.A. Gunderson, N.T. Edwards, and D.W. Johnson. (2002). Net primary productivity of a $CO_2$–enriched deciduous forest and the implications for carbon storage. *Ecol. Appl.* 12, 1261–1266.

Novara, A., Armstrong, A., Gristina, L., Semple, K.T., and Quinton, J.N. 2012. Effects of soil compaction, rain exposure and their interaction on soil carbon dioxide emission. *Earth Surf. Process. Landf.* 37, 994–999. doi:10.1002/esp.3224

O'Neill, K.P., Amacher, M.C., and Palmer, C.J. (2005a). Developing a national indicator of soil quality on U.S. forestlands: Methods and initial results. *Environ. Monit. Assess.* 107, 59–80. doi:10.1007/s10661-005-2144-0

O'Neill, K.P., Amacher, M.C., and Perry, C.H. (2005b). Soils as an indicator of forest health: A guide to the collection, analysis, and interpretation of soil indicator data in the Forest Inventory and Analysis program. Gen. Tech. Rep. NC-258. St. Paul, MN: U.S. Department of Agriculture Forest Service, Northern Research Station.

Ostertag, R., Scatena, F.N., and Silver, W.L. (2003). Forest floor decomposition following hurricane litter inputs in several Puerto Rican forests. *Ecosystems* 6, 261–273. doi:10.1007/PL00021512

Page-Dumroese, D.S., Brown, R.E., Jurgensen, M.F., and Mroz, G.D. (1999). Comparison of methods for determining bulk densities of rocky forest soils. *Soil Sci. Soc. Am. J.* 63, 379–383. doi:10.2136/sssaj1999.03615995006300020016x

Page-Dumroese, D.S., Abbott, A.M., and Rice, T.M. (2009). *Forest floor disturbance monitoring protocol Volume II: Supplementary methods, statistics, and data collection.* Gen. Tech. Rep. WO-82b. Fort Collins, CO: U.S. Department of Agriculture, Forest Service, Rocky Mountain Research Station.

Page-Dumroese, D.S., Jurgensen, M., and Terry, T. (2010). Maintaining soil productivity during forest or biomass-to-energy thinning harvests in the western United States. *West. J. Appl. For.* 25, 5–11. doi:10.1093/wjaf/25.1.5

Page-Dumroese, D.S., Busse, M.D., Overby, S.T., Gardner, B.D., and Tirocke, J.M. (2015). Impacts of forest harvest on active carbon and microbial properties of a volcanic ash cap soil in northern Idaho. *J. Soil Sci.* 5, 11–19.

Page-Dumroese, D.S., Coleman, M.D., and Thomas, S.C. (2018). Opportunities and uses of biochar on forest sites in North America. In: V.J. Bruckman, E.A. Varol,

B.B. Uzun, and J. Liu, (eds.), *Biochar: A regional supply chain approach in view of climate change mitigation* (p. 315–336). Cambridge, U.K.: Cambridge Univ. Press.

Paoletti, M.G. (1999). The role of earthworms for assessment sustainability and as bioindicators. *Agric. Ecosyst. Environ.* 74, 137–155. doi:10.1016/S0167-8809(99)00034-1

Paudel, B.R., Udawatta, R.P., Kremer, R.J., and Anderson, S.H. (2011). Agroforestry and grass buffer effects on soil quality parameters for grazed pasture and row-crop systems. *Appl. Soil Ecol.* 48, 125–132. doi:10.1016/j.apsoil.2011.04.004

Paudel, B.R., Udawatta, R.P., Kremer, R.J., and Anderson, S.H. (2012). Soil quality indicator responses to crop, grazed pasture, and agroforestry buffer management. *Agrofor. Syst.* 84, 311–323. doi:10.1007/s10457-011-9454-8

Paz-Ferreiro, J., Lu, H., Fu, S., Méndez, A., and Gascó, G. (2014). Use of phytoremediation and biochar to remediate heavy metal polluted soils: A review. *Solid Earth* 5, 65–75. doi:10.5194/se-5-65-2014

Pellegrini, A.F.A., Ahlström, A., Hobbie, S.E., Reich, P.B., Nierazdik, L.P., Staver, A.C., Scharenbroch, B.C., Jumpponen, A., Anderegg, W.R.L., Randerson, J.T., and Jackson, R.B. (2017). Fire frequency drives decadal changes in soil carbon and nitrogen and ecosystem productivity. *Nature* 553, 194–198. doi:10.1038/nature24668

Perry, C.H., and Amacher, M.C. (2007a). Chemical properties of forest soils. In M.J. Ambrose and B.L. Conkling, (eds.), *Forest health monitoring: 2005 National Technical Report* (p. 59–66). Gen. Tech. Rep. SRS-GTR-104. Asheville, NC: U.S. Department of Agriculture, Forest Service, Southern Research Station.

Perry, C.H., and Amacher, M.C. (2007b). Physical properties of forest soils. In M.J. Ambrose and B. L. Conkling (eds.), *Forest health Monitoring: 2005 National Technical Report* (p. 51–58). Gen. Tech. Rep. SRS-GTR-104. Asheville, NC: U.S. Department of Agriculture, Forest Service, Southern Research Station.

Perry, C.H., and Amacher, M.C. (2007c). Soil carbon. In M.J. Ambrose and B.L. Conkling (eds.), *Forest health monitoring: 2005 National Technical Report* (p. 67–72). Gen. Tech. Rep. SRS-GTR-204. Asheville, NC: U.S. Department of Agriculture, Forest Service, Southern Research Station.

Perry, C.H., and Amacher, M.C. (2009). Forest soils. In W.B. Smith et al. (eds). *Forest resources of the United States, 2007.* Gen. Tech. Rep. WO-GTR-78. Washington, D.C.: U.S. Department of Agriculture, Forest Service.

Perry, C.H., and Amacher, M.C. (2012). Patterns of soil calcium and aluminum across the conterminous United States. In K.M. Potter and B.L. Conkling (eds). *Forest health monitoring: 2008 National Technical Report* (p. 119-130). Gen. Tech. Rep. SRS-GTR-158. Asheville, NC: U.S. Department of Agriculture Forest Service, Southern Research Station.

Perry, C.H., Woodall, C.W., Amacher, M.C., and O'Neill, K.P. (2009). An inventory of carbon storage in forest soil and down woody material of the United States. In: B.J. McPherson and E.T. Sundquist, editors, *Carbon sequestration and its role in the global carbon cycle* (p. 101–116). Washington, D.C.: American Geophysical Union. doi:10.1029/2006GM000341

Perry, C.H., and P.L. Zimmerman. 2012. Building improved models of sugar maple mortality. In R.S. Morin and G.C. Liknes, (eds.), *Moving from status to trends: Forest Inventory and Analysis (FIA) symposium 2012* (p. 204–209). Proceedings. NRS-P-105. Baltimore, MD: U.S. Department of Agriculture, Forest Service, Northern Research Station.

Phillips, R.P., I.C. Meier, E.S. Bernhardt, A.S. Grandy, K. Wickings, and A.C. Finzi. (2012). Roots and fungi accelerate carbon and nitrogen cycling in forests exposed to elevated $CO_2$. *Ecol. Lett.* 15, 1042–1049. doi:10.1111/j.1461-0248.2012.01827.x

Pinho, R.C., Miller, R.P., and Alfaia, S.S. (2012). Agroforestry and the improvement of soil fertility: A view from Amazonia. *App. Environ. Soil Sci.* 212, 616383.

Polglase, P., Paul, K., Hawkins, C., Siggins, A., Turner, J., Booth, T., Crawford, D, Jovanovic, T., Hobbs, T., Opie, K., Almeida, A., and Carter, J. 2008. Regional opportunities for agroforestry systems in Australia. RIRDC/L and WA/FWPA/MDBC Joint Venture Agroforestry Program. Kingston, Australia: RIRDC.

Powers, R.F., D.A. Scott, F.G. Sanchez, R.A. Voldseth, D. Page-Dumroese, J.D. Elioff, and D.M. Stone. 2005. The North American long-term soil productivity experiment: Findings from the first decade of research. *For. Ecol. Manage.* 220, 31–50. doi:10.1016/j.foreco.2005.08.003

Reams, G.A., Smith, W.D., Hansen, M.H., Bechtold, W.A., Roesch, F.A., and Moisen, G.G. (2005). The Forest Inventory and Analysis sampling frame. In W.A. Bechtold and P. L. Patterson (eds), *The enhanced Forest Inventory and Analysis program—national sampling design and estimation procedures* (p. 11–26). Gen. Tech. Rep. SRS-GTR-80. Asheville, NC: U.S. Department of Agriculture Forest Service, Southern Research Station.

Reeves, D.A., Reeves, M.C., Abbott, A.M., Page-Dumroese, D.S., and Coleman, M.D. (2012). A detrimental soil disturbance prediction model for ground-based timber harvesting. *Can. J. For. Res.* 42, 821–830. doi:10.1139/x2012-034

Richardson, B.A., Richardson, M.J., González, G., Shiels, A.B., and Srivastava, D.S. (2010). A canopy trimming experiment in Puerto Rico: The response of litter invertebrate communities to canopy loss and debris deposition in a tropical forest subject to hurricanes. *Ecosystems* 13, 286–301. doi:10.1007/s10021-010-9317-6

Richter, D.D., and Markewitz, D. (2001). *Understanding soil change: Soil sutainability over millennia, centuries, and decades.* New York: Cambridge Univ. Press.

Richter, D.D., Markewitz, D., Trumbore, S.E., and Wells, C.G. (1999). Rapid accumulation and turnover of soil carbon in a re-establishing forest. *Nature* 400, 56–58. doi:10.1038/21867

Rivest, D., M. Lorente, A. Oliver, and Messier, C. (2013). Soil biochemical properties and microbial resilience in agroforestry systems: Effects on wheat growth under controlled drought and flooding conditions. *Sci. Total Environ.* 463–464, 51–60. doi:10.1016/j.scitotenv.2013.05.071

Rohr, T., Manzoni, S., Feng, X., Menezes, R.S., and Porporato, A. (2013). Effect of rainfall seasonality on carbon storage in tropical dry ecosystems. *J. Geophys. Res. Biogeosci.* 118, 1156–1167. doi:10.1002/jgrg.20091

Ross-Davis, A.L., J.E. Stewart, and M. Settles., J.W. Hanna [et al.] 2016. Fine-scale variability of forest soil fungal communities in two contrasting habitat series in northern Idaho, USA. p. 145–149 In A. Ramsey and P. Palacios, *Proceedings of the 63rd annual Western International Forest Disease Work Conference*, 21-25 September 2015, Newport, OR. Fort Collins, CO: Western Interntional Forest Disease Work Conference.

Sanchez, F.G., Tiarks, A.E., Kranabetter, J.M., Page-Dumroese, D.S., Powers, R.F., Sanborn, R.T., Chapman, W.K. (2006a). Effects of organic matter removal and soil compaction on fifth-year mineral soil carbon and nitrogen contents for sites across the United States and Canada. *Can. J. For. Res.* 36(3), 565–576. doi:10.1139/x05-259

Sanchez, F.G., Scott, D.A., and Ludovici, K.H. (2006b). Negligible effects of severe organic matter removal and soil compaction on loblolly pine growth over 10 years. *For. Ecol. Manage.* 227, 145–154. doi:10.1016/j.foreco.2006.02.015

Sánchez-DeLeon, Y., Zou, X., Borges, S., and. Ruan, H.H. (2003). Recovery of native earthworms in abandoned tropical pastures. *Conserv. Biol.* 17, 999–1006. doi:10.1046/j.1523-1739.2003.02098.x

Sánchez-DeLeon, Y., and Zou, X. (2004). Plant influences on native and exotic earthworms during secondary succession in old tropical pastures. *Pedobiologia* 48, 215–226. doi:10.1016/j.pedobi.2003.12.006

Sanford, R.L., Jr., Parton, W.J., Ojima, D.S., and Lodge, D.J. (1991). Hurricane effects on soil organic matter dynamics and forest production in the Luquillo Experimental Forest, Puerto Rico: Results of simulation modeling. *Biotropica* 23, 364–372. doi:10.2307/2388253

Sauer, T.J., Cambardella, C.A., and Brandle, J.R. (2007). Soil carbon and litter dynamics in a red cedar-scotch pine shelterbelt. *Agrofor. Syst.* 71, 163–174. doi:10.1007/s10457-007-9072-7

Scatena, F.N., and Lugo, A.E. (1995). Geomorphology, disturbance, and the soil and vegetation of 2 subtropical wet steepland watersheds of Puerto Rico. *Geomorphology* 13, 199–213. doi:10.1016/0169-555X(95)00021-V

Schloter, M., Dilly, O., and Munch, J.C. (2003). Indicators for evaluating soil quality. *Agric. Ecosyst. Environ.* 98, 255–262. doi:10.1016/S0167-8809(03)00085-9

Schoeneberger, M., Bentrup, G., de Gooijer, H., Soolanayakanahally, R., Sauer, T., Brandle, J., Zhou, X., and Current, D. 2012a. Branchinhg out: Agroforestry as a climate change mitigation and adaptation tool for agriculture. *J. Soil Water Conserv.* 67:128A–136A. doi:10.2489/jswc.67.5.128A

Schoeneberger, P.J., Wysocki, D.A., Benham, E.C., and Soil Survey Staff. (2012b). *Field book for describing and sampling soils, Version 3.0*. Lincoln, NE: U.S. Department of Agriculture, Natural Resources Conservation Service, National Soil Survey Center.

Schoenholtz, S.H., Van Miegroet, H., and Burger, J.A. (2000). A review of chemical and physical properties as indicators of forest soil quality: Challenges and opportunities. *For. Ecol. Manage.* 138, 335–356. doi:10.1016/S0378-1127(00)00423-0

Schroeder, P. (1993). Agroforestry systems: Integrated land use to store and conserve carbon. *Climate Research Special: Terrestrial Management* 3, 53–60. doi:10.3354/cr003053

Schultz, R.C., Isenhart, T.M., Colletti, J.P., Simpkins, W.W., Udawatta, R.P., and Schultz, P.L. (2009). Riparian and upland buffer practices. In: H.E. Garrett, editor, *North American agroforestry: An integrated science and practice* (p. 163–218). 2nd ed. Madison, WI: ASA.

Scott, C.T., Bechtold, W.A., Reams, G.A., and Smith, W.D. (2005). Sample-based estimators used by the Forest Inventory and Analysis National Information Management System. In W.A. Bechtold and P.L. Patterson, editors, The Enhanced Forest Inventory and Analysis Program: National sampling design and estimation procedures (p. 43–67). Gen. Tech. Rep. SRS-GTR-80. Asheville, NC: U.S. Department of Agriculture Forest Service, Southern Research Station.

Seiter, S., Ingham, E.R., William, R.D., and Hibbs, D.E. (1995). Increase in soil microbial biomass and transfer of nitrogen from alder to sweet corn in an alley cropping system. In J.H. Ehrenreich, et al., editors, *Growing a sustainable future* (p. 56–158). Boise, ID: University of Idaho.

Senyk, J.P., and Craigdallie, D. (1997). *Effects of harvesting methods on soil properties and forest productivity in interior British Columbia*. Victoria, BC: Canadian Forest Service, Pacific Forestry Center IX.

Seobi, T., Anderson, S.H., Udawatta, R.P., and Gantzer, C.J. (2005). Influences of grass and agroforestry buffer strips on soil hydraulic properties. *Soil Sci. Soc. Am. J.* 69:893–901. doi:10.2136/sssaj2004.0280

Sharrow, S.H., and Ismail, S. (2004). Carbon and nitrogen storage in agroforests, tree plantations, and pastures in western Oregon, USA. *Agrofor. Syst.* 60:123–130. doi:10.1023/B:AGFO.0000013267.87896.41

Shiels, A.B., and González, G. (2014). Understanding the key mechanisms of tropical forest responses to canopy loss and biomass deposition from experimental hurricane effects. *For. Ecol. Manage.* 332, 1–10. doi:10.1016/j.foreco.2014.04.024

Shiels, A.B., González, G., Lodge, D.J., Willig, M.R., and Zimmerman, J.K. (2015). Cascading effects of canopy opening and debris deposition from a large-scale hurricane experiment in a tropical rain forest. *BioSci.* 65, 871–881. doi:10.1093/biosci/biv111

Sigua, G.C. (2018). Effects of crop rotations and intercropping on soil health. In D. Reicosky, editor, *Managing soil health for sustainable agriculture* (p. 163–189). Sawston, UK: Burleigh Dodds Science Publishing. doi:10.19103/AS.2017.0033.25

Silver, W.L. (1998). The potential effects of elevated $CO_2$ and climate change on tropical forest soils and biogeochemical cycling. *Clim. Change* 39, 337–361. doi:10.1023/A:1005396714941

Six, J., and Jastrow, J.D. (2002). Organic matter turnover. In R. Lal, (ed.), *Encyclopedia of soil science* (p. 936–942). New York: Marcel Dekker, Inc.

Shestak, C.J., and Busse, M.D. (2005). Compaction alters physical but not biological indices of soil health. *Soil Sci. Soc. Am. J.* 69(1), 236–246. doi:10.2136/sssaj2005.0236

Slesak, R.A., Palik, B.J., D'Amato, A.W., and Kurth, V.J. (2017). Changes in soil physical and chemical properties following organic matter removal and compaction: 20-year response of the aspen Lake-States Long-Term Soil Productivity installations. *For. Ecol. Manage.* 392, 68–77. doi:10.1016/j.foreco.2017.03.005

Smith, J.E., and Heath, L.S. (2002). *A model of forest floor carbon mass for United States forest types*. Res. Pap. NE-RP-722. Newtown Square, PA: U.S. Department of Agriculture Forest Service, Northeastern Research Station. doi:10.2737/NE-RP-722

Smith, P., Martono, D., Cai, Z., Gwary, D., and Janzen, H. (2007). Agriculture. In: B. Metz, et al., editors, *Climate change 2007: Mitigation* (p. 497–540). Cambridge, UK: Cambridge Univ. Press.

Soil Science Division Staff. 2017. Soil survey manual. In C. Ditzler, K. Scheffe, & H.C. Monger, *USDA Handbook 18*. Washington, DC: Government Printing Office.

Spears, J.D.H., Holub, S.M., and Lajtha, K. (2003). The influence of decomposing logs on soil biology and nutrient cycling in an old-growth mixed coniferous forest in Oregon, U.S.A. *Can. J. For. Res.* 33, 2193–2201. doi:10.1139/x03-148

Spillers, A.R. (1939). *Forest resources of southeast Alabama*. Forest Survey Release No. 47. New Orleans, LA: U.S. Department of Agriculture Forest Service, Southern Forest Experiment Station.

Stefano, A.D., and Jacobson, M.G. (2018). Soil carbon sequestration in agroforestry systems: A meta-analysis. *Agrofor. Syst.* 92, 285–299.

Sudduth, K.A., Myers, D.B., Kitchen, N.R., and Drummond, S.T. (2013). Modeling soil electrical conductivity-depth relationships with data from proximal and penetrating ECa sensors. *Geoderma* 199, 12–21. doi:10.1016/j.geoderma.2012.10.006

Swift, M.J., Heal, O.W., and Anderson, J.W. (1979). *Decomposition in terrestrial ecosystems*. Blackwell: Oxford, UK.

Tringe, S. G., C. Von Mering, A. Kobayashi, A. A. Salamov, K. Chen, H.W. Chang, M. Podar, J.M. Short, E.J. Mathur, J.C. Detter, P. Bork, P. Hugenholtz, and E.M. Rubin. (2005). Comparative metagenomics of microbial communities. *Science* 308, 554–557. doi:10.1126/science.1107851

Torres, J.A., and González, G. (2005). Wood decomposition in Puerto Rican dry and wet forests: A 13-year case study. *Biotropica* 37(3), 452–456. doi:10.1111/j.1744-7429.2005.00059.x

Udawatta, R.P., and Jose, S. (2012). Agroforestry strategies to sequester carbon in temperate North America. *Agrofor. Syst.* 86, 225–242. doi:10.1007/s10457-012-9561-1

Udawatta, R.P., Anderson, S.H., Gantzer, C.J., and Garrett, H.E. (2006). Agroforestry and grass buffer influence on macropore characteristics: A computed tomography analysis. *Soil Sci. Soc. Am. J.* 70, 1763–1773. doi:10.2136/sssaj2006.0307

Udawatta, R.P., Gantzer, C.J., Anderson, S.H., and Garrett, H.E. (2008). Agroforestry and grass buffer effects on high resolution X-ray CT-measured pore characteristics. *Soil Sci. Soc. Am. J.* 72, 295–304. doi:10.2136/sssaj2007.0057

Udawatta, R.P., Kremer, R.J., Garrett, H.E., and Anderson, S.H. (2009). Soil enzyme activities and physical properties in a watershed managed under agroforestry and row-crop system. *Agric. Ecosyst. Environ.* 131, 98–104. doi:10.1016/j.agee.2008.06.001

Udawatta, R.P., Anderson, S.H., Motavalli, P.P., and Garrett, H.E. (2011a). Clay and temperature influences on sensor measured volumetric soil water content. *Agrofor. Syst.* 82, 61–75. doi:10.1007/s10457-010-9362-3

Udawatta, R.P., Garrett, H.E., and Kallenbach, R.L. (2011b). Agroforestry buffers for non-point source pollution reductions from agricultural watersheds. *J. Environ. Qual.* 40, 800–806. doi:10.2134/jeq2010.0168

US Environmental Protection Agency (EPA). (2018). *Inventory of U.S. greenhouse gas emissions and sinks: 1990–2016. EPA 430-R-18-003*. Washington, D.C.: US Environmental Protection Agency.

USDA Forest Service. (2011). *Phase 3 field guide: Soil measurements and sampling, version 5.1*. www.fia.fs.fed.us/library/field-guides-methods-proc/ (Accessed 29 Aug. 2018).

USDA Forest Service. (2017). *Forest inventory and analysis national core field guide. Volume 1: Field data collection procedures for phase 2 plots, version 7.2*. www.fia.fs.fed.us/library/field-guides-methods-proc/ (Access 29 Aug 2018).

van Bruggen, A.H.C., and Semenov, A.M. (2000). In: search of biological indicators for soil health and disease suppression. *Appl. Soil Ecol.* 15, 13–24. doi:10.1016/S0929-1393(00)00068-8

Van Doren, C.A., and A.A. Klingebiel. (1952). Effect of management on soil permeability. *Soil Science Society of America Proceedings* 15, 399–408.

van der Heijden, M.G.A., Bardgett, R.D., and Van Straalen, N.M. (2008) The unseen majority: Soil microbes as drivers of plant diversity and productivity in terrestrial ecosystems. *Ecol. Lett.* 11(3), 296–310. doi:10.1111/j.1461-0248.2007.01139.x

van Straalen, N.M. (1998). Evaluation of bioindicator systems derived from soil arthropod communities. *Appl. Soil Ecol.* 9, 429–437. doi:10.1016/S0929-1393(98)00101-2

Wang, Q.K., and Wang, S.L. (2007). Soil organic matter under different forest types in Southern China. *Geoderma* 142, 349–356. doi:10.1016/j.geoderma.2007.09.006

Watson, R.T., Noble, I.R., Bolin, B., Ravindranathan, N.R., Verardo, D.J., and Dokken, D.J. (2000). *IPCC special report on land use, land use change, and forestry.* Cambridge, U.K.: Cambridge University Press. http://www.rida.no/climate/ipcc/land_use/

Wagenet, R.J., and Hutson, J.L. (1997). Soil quality and its dependence on dynamic physical processes. *J. Environ. Qual.* 26, 41–48. doi:10.2134/jeq1997.00472425002600010007x

Wilson, B.T., Lister, A.J., and Riemann, R.I. (2012). A nearest-neighbor imputation approach to mapping tree species over large areas using forest inventory plots and moderate resolution raster data. *For. Ecol. Manage.* 271,182–198. doi:10.1016/j.foreco.2012.02.002

Wilson, B.T., Woodall, C.W., and Griffith, D.M. (2013). Imputing forest carbon stock estimates from inventory plots to a nationally continuous coverage. *Carbon Balance Manage.* 8, 1. http://www.cbmjournal.com/content/8/1/1/ doi:10.1186/1750-0680-8-1

Woodall, C. W., Amacher, M.C., Bechtold, W.A., Coulston, J.W., Jovan, S., Perry, C.H., Randolph, K.C., Schulz, B.K., Smith, G.C., Tkacz, B., and Will-Wolf, S. (2011). Status and future of the forest health indicators program of the United States. *Environ. Monit. Assess.* 177, 419–436. doi:10.1007/s10661-010-1644-8

Zalamea, M., González, G., Ping, C.L., and Michaelson, G. (2007). Soil organic matter dynamics under decaying wood. *Plant Soil* 296, 173–185. doi:10.1007/s11104-007-9307-4

Zalamea, M., González, G., and Lodge, D.J. (2016). Physical, chemical, and biological properties of soil under decaying wood in a tropical wet forest in Puerto Rico. *Forests* 7(8), 168. doi:10.3390/f7080168

Zalesny, R.S., Headlee, W.L., Gopalakrishnan, G., Bauer, E.O., Hall, R.B., Hazel, D.W., Isebrands, J.G., Licht, L.A., Negri, M.C., Nichols, E.G., Rockwood, D.L., Wiese, A.H. (2019). Ecosystem services of poplar at long-term phytoremediation sites in the Midwest and Southeast, United States. *Wiley Interdisciplinary Reviews: Energy and Environment*. 8(6), E349.

Zimmerman, J.K., Pulliam, W.M., Lodge, D.L., Quiñones-Orfila, V, Fletcher, N., Guzmán-Grajales, S., Parotta, J.A., Asbury, C.E., Walker, L.R., and R.B. Wade. (1995). Nitrogen immobilization by decomposing woody debris and the recovery of tropical wet forest from hurricane damage. *Oikos* 72, 314–322. doi:10.2307/3546116

Zomer, R.J., Trabucco, A., Coe, R., and Place, F. (2009). *Trees on farm: Analysis of global extent and geographical patterns of agroforestry*. ICRAF Working Paper no. 89. Nairobi, Kenya: World Agroforestry Centre.

Zou, X., and González, G. (1997). Changes in earthworm density and community structure during secondary succession in abandoned tropical pastures. *Soil Biol. Biochem.* 29(3-4), 627–629. doi:10.1016/S0038-0717(96)00188-5

Zou, X., and González, G. (2001). Earthworms in tropical tree plantations: Effects of management and relations with soil carbon and nutrient use efficiency. In M.V. Reddy, editor, *Management of tropical plantation-forests and their soil litter system: Litter, biota and soil nutrient dynamics* (p. 283–295). Enfield, NH: Science Publishers, Inc.

# 7

# A Risk-Based Soil Health Approach to Management of Soil Lead

*Nicholas T. Basta, Alyssa M. Zearley, Jeffory A. Hattey, and Douglas L. Karlen*

## Introduction

Lead (Pb) is one of the most common urban legacy soil contaminants due to mining, refining and industrial processes, historical use in plumbing systems, and ubiquitous inclusion in gasoline and paint products throughout much of the 20th Century. These anthropogenic activities have elevated soil Pb to levels that can potentially impact both human and soil health. This chapter provides an overview of legacy soil Pb sources, exposure pathways, risk assessment, and restoration strategies to improve soil health and reduce risk to human health.

Low, background concentrations of soil Pb are associated with minerals such as anglesite, cerussite, and galena. Typical quantities have been documented in two nationwide surveys conducted by the United States Geological Survey (USGS) which report which show natural mean levels of background soil Pb as 19 mg Pb $kg^{-1}$ soil (mean) (Shacklette & Boerngen, 1984), 16.5 mg Pb $kg^{-1}$ soil (median), and 22 mg Pb $kg^{-1}$ soil (mean) (Smith et al., 2013).

The source of anthropogenic Pb affects location and concentration in excess of background levels. When associated with mining, refining and industrial processes Pb deposition is more localized, often with very high levels of soil Pb. Concentrations in those soils may reach tens of thousands mg Pb $kg^{-1}$ soil, much higher than the United States Environmental Protection Agency (USEPA)

*Soil Health Series: Volume 1 Approaches to Soil Health Analysis*, First Edition.
Edited by Douglas L. Karlen, Diane E. Stott, and Maysoon M. Mikha.

residential soil screening limit of 400 mg Pb $kg^{-1}$ soil, and definitely posing considerable risk to human health (USEPA 1994, 2016).

Lead use in gasoline, paint and plumbing products along with smaller industrial activities like "mom and pop" smelters and battery recycling (USA Today, 2012), dispersed lower concentrations of Pb over larger areas (Mielke, 2019). It is estimated these emissions deposited 10 million tons of Pb in urban soil, water, and air environments (Johnson et al., 2012). These sources were phased out in the United States in the late 20th century by a series of federal regulations. These positive regulatory actions resulted in a corresponding decrease of human blood lead levels (BLL) (Figure 7.1).

Although environmental emissions have been significantly reduced, Pb in soil is persistent and legacy pollution has remained for decades (Minca & Basta, 2013). Also, soil accumulation of Pb continues to occur from Pb-based paint or repurposing smelter waste as a general "clean fill". Such actions have resulted in off-site soil contamination of parks, residences and surrounding communities (Kansas Department of Health and Environment, 2019), thus continuing to pose a human health risk, with Pb ingestion as the primary pathway (USEPA, 2002).

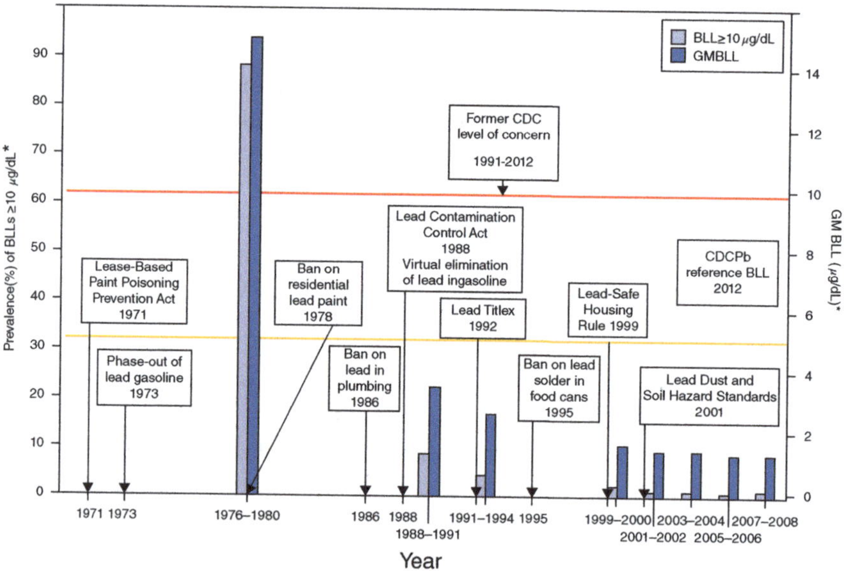


**Figure 7.1** Timeline of Geometric mean (GM) and excessive (>10 ug/dL) blood lead level (BLL) in U.S. children (1 to 5 yr of age) associated with lead poisoning prevention policies. Data from the 1971 to 2008 National Health and Nutrition Examination Survey; adapted from Centers for Disease Control and Prevention (CDC) publication by Brown and Margolis (2012).

Soil health impacts from Pb contamination include: (i) increases and decreases in soil microbial populations (Doelman & Haanstra, 1979b; Konopka et al., 1999; Liao et al., 2007; Shi et al., 2002a; Shi et al., 2002b), (ii) decreased microbial biomass (Zeng et al., 2006), (iii) diminished soil respiration (Doelman & Haanstra, 1979a, 1984; Zeng et al., 2006), (iv) decreased protein formation (Shi et al., 2002a, 2002b), and (v) lower enzyme activity (Doelman & Haanstra, 1979a; Marzadori et al., 1996; Murata et al., 2005). While it is clear that Pb can alter soil microbial community composition and functioning, implications for soil health are less studied. A few studies show that selection of Pb tolerant species, with reduced physiological activity, can change a soil's capacity to adapt to ecosystem stress. Soil Pb can reduce nutrient availability to plants because Pb does inhibit nitrification (Li et al., 2016). Other studies (e.g., Suhadolc et al., 2004) found herbicides were degraded more slowly in Pb-spiked soils, suggesting Pb may influence breakdown of some organic compounds and pollutants. Similarly, Pb acted synergistically with the antibiotic oxytetracycline and reduced sucrase activity in a pot study (Gao et al., 2013). Soil Pb is often associated with other soil contaminants, interactions among them are virtually unstudied.

Lead levels in soil directly link soil health and human health, as exposure to Pb is associated with numerous negative health effects. Toxic to all organ systems, Pb is particularly damaging to the kidneys, cardiovascular system, and nervous system (Schroder et al., 2004). Childhood exposure to Pb during key developmental periods can result in permanent neurological impairment. The U.S. Center for Disease Control and Prevention (CDC) Advisory Committee on Childhood Lead Poisoning Prevention (ACCLPP) defined "elevated" blood lead (EBL), which necessitates follow-up investigation, as 5 μg $dL^{-1}$, the 95th percentile of blood lead distribution in 2012. Neurological outcomes associated with children blood Pb above EBL includes a 2 to 4 point IQ deficit for each μg $dL^{-1}$ in the range of 5 to 35 μg $dL^{-1}$). Approximately 500,000 children in the U.S. exceeded the BLL of 5 μg $dL^{-1}$ in 2017 (Hauptman et al., 2017). Older cities tend to have higher rates of excessive blood lead level (BLL) in children due in part to higher soil Pb concentrations. In 2017, more than 12% of tested children had elevated BLL in Cleveland, Ohio (Figure 7.2; Cuyahoga County Board of Health, 2017). In the same year, Detroit, found that 9% of children under 6 tested had elevated BLLs (Michigan Department of Health & Human Services, 2018).

Numerous studies (Datko-Williams et al., 2014; Duzgoren-Aydin et al., 2006; Mielke et al., 2011; Kelly et al., 1996; Minca et al., 2013) have shown soil Pb concentrations are higher in urban areas than suburban or rural areas, in larger cities than smaller cities, and in older cities than younger cities. A detailed distribution of Pb across the U.S. by Mielke et al. (2011) and Datko-Williams et al. (2014) shows soil Pb is more prevalent in urban areas due to denser traffic, industry, and

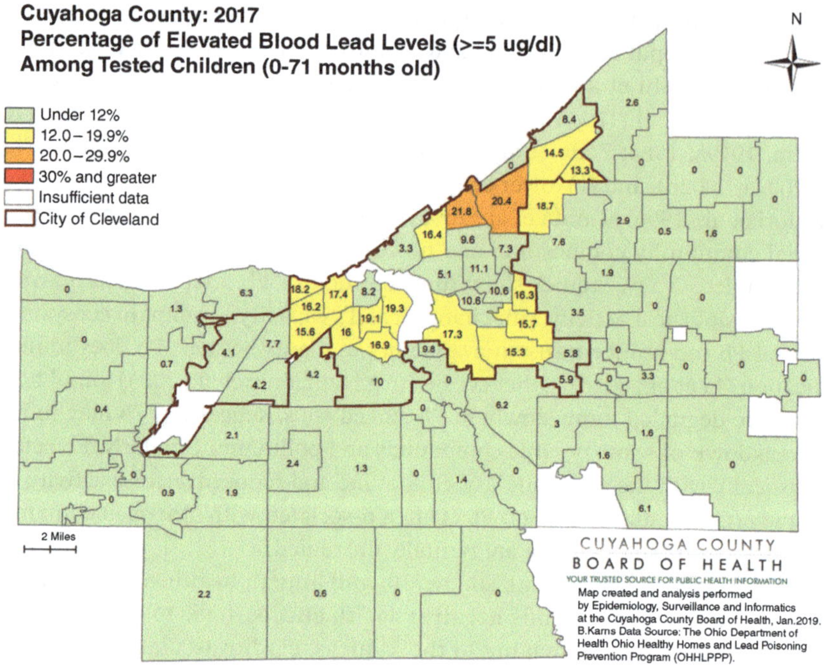


**Figure 7.2** Blood Lead Levels of children 0–71 months in Cuyahoga County, OH, 2016.

housing. In the City Park neighborhood of Appleton, Wisconsin, 40% of yard space had greater than 400 mg Pb $kg^{-1}$ soil, the USEPA screening limit for residential soil (Clark & Knudsen, 2013). Soils sampled from around the Hough neighborhood in Cleveland, OH found 25% exceeded 400 mg Pb $kg^{-1}$ soil (Sharma et al., 2015). Reducing exposure to soil Pb should be a priority in these areas. Although the science of soil Pb contamination is extensive, soil health assessment and restoration of Pb-contaminated urban soil is lacking. The objective of this chapter is to provide a risk-based theoretical and practical foundation for a soil health approach to management of urban soil Pb.

## Urban Soil Lead Assessment and Human Exposure

Human exposure to Pb occurs in both built and urban environments. In built environments (homes, schools, daycare facilities, etc.), interior Pb-based paint is a significant contributor to exposure that has been documented in other reviews (e.g., Hauptman et al., 2017). In urban environments, soil exposure is a significant contributor to Pb in children because they have higher rates of incidental soil ingestion than adults (Mielke et al., 2019; Mielke et al., 1999). Direct or incidental ingestion of Pb contaminated soil or dust is often associated with consumption of homegrown produce, as recognized in several studies of children with elevated BLL (Farrell et al., 1998; Mielke, 1999; Mielke et al., 1992; Mielke et al., 2019; Mielke et al., 2011; Weitzman et al., 1993). The primary mechanisms of exposure to soil Pb occur through ingestion of dust particles transferred from hands or cleared from the airway (Chaney & Ryan, 1994), as inhalation of dust particles ($PM_{10}$) and dermal absorption are minor exposure pathways.

Gardeners face additional exposure to soil Pb through incidental ingestion of dust while working in the garden, hand-to-mouth transfer, and consumption of contaminated vegetables (Brown et al., 2016; Khan et al., 2015; McBride et al., 2014; Preer et al., 1984), although research quantifying exposure from those pathways is limited and findings are conflicting. Spliethoff et al. (2016) reported that incidental soil ingestion was the largest Pb contributor for children, while consumption of garden produce was the primary risk for adults. Conversely, Hough et al. (2004) concluded that Pb-contaminated produce was the most significant Pb source for gardeners of all ages.

Vegetables provide Pb exposure by metabolic uptake in root or foliar tissue with transfer to fruit and external adhesion of soil particles to roots, leaves or fruit (McBride et al., 2014; Olowoyo & Lion, 2016). The World Health Organization (WHO) and Food and Agriculture Organization of the United Nations (FAO) have set maximum allowable concentrations for Pb in vegetables at 0.1 to 0.3 mg Pb $kg^{-1}$ fresh tissue weight (WHO/FAO, 2015). Crops grown in rural agricultural soils with background Pb concentrations rarely exceed these limits, but vegetables grown in Pb-contaminated soils routinely exceed these Pb concentration limits. McBride et al. (2014) found 9% of leafy green vegetables and 47% of root crops from New York City gardens exceeded the WHO/FAO guidelines for maximum Pb concentrations by up to four times. Studies generally find that vegetable fruits have the lowest tissue Pb concentrations while leafy vegetables and true root crops have higher tissue Pb concentrations (Fan et al., 2017; Finster et al., 2004; McBride et al., 2014).

Soil Pb concentration is not a good predictor of vegetable Pb content. Transfer of Pb from soil to vegetables depends on numerous factors including: soil Pb concentration and form, vegetable species and variety (Alexander et al., 2006), plant part (Finster et al., 2004), soil pH, soil organic matter (SOM), soil nutrients, properties such as texture and erosivity that influence soil adhesion to vegetable surfaces, and aerial deposition. The lack of correlation between soil and vegetable Pb concentrations was confirmed by McBride et al. (2014) who found no relationship in New York City gardens and by Finster et al. (2004) who collected similar data in Chicago. Lead tansfer factors were reviewed by Khan et al. (2015) and varied from 0.002 to 0.25 and 1.26 to 14 mg Pb $kg^{-1}$ tissue/mg Pb $kg^{-1}$ soil for lettuce and tomatoes, respectively. Transfer mechanisms varied by a factor of 100 for the same variety of cabbage in a pot study (Alloway et al., 1988). Vegetables can exceed Pb concentration limits even when grown on relatively low Pb soils. Urban garden soils in Nigeria, with average Pb concentrations below 50 mg Pb $kg^{-1}$ soil, produced lettuce with greater than 2 mg Pb $kg^{-1}$ fresh tissue weight, which is over six times the WHO/FAO limits (Agbenin et al., 2009). On the other hand, vegetables grown in highly contaminated soil may have Pb concentrations below WHO/FAO limits (Spittler & Feder, 1979).

Ingestion of plant material containing Pb does not infer 100% relative bioavailability (RBA) for humans, although it is currently considered to be 100% RBA in human health risk assessment. Consuming food products concurrently with high Pb concentrations can significantly reduce bioavailability or the portion of Pb absorbed into the bloodstream (Barltrop & Khoo, 1975; Mylroie et al., 1978; Wise & Gilburt, 1980). Maddaloni et al. (1998) found that subjects absorbed 26% of the lead in soil ingested while fasting, but a different group, dosed with Pb-contaminated soil during a breakfast meal, absorbed only 2.5% of the dosed Pb. James et al. (1985) reported similar results when Pb was administered following meals. Given that most food items can reduce soil Pb bioavailability (Deluca et al., 1982; Fu & Cui, 2013; James et al., 1985; Yang et al., 2012), it may be that Pb consumed with urban-grown vegetables is not 100% bioavailable, but again, the impact of vegetables on Pb bioavailability has not been extensively evaluated. A study by H. B. Li et al. (2018) showed mice fed Pb-contaminated soil with apples or cabbage absorbed less Pb than mice fed soil while fasting. Vegetables have also been shown to reduce soil Pb bioaccessibility using gastrointestinal in vitro methods. Using the physiologically based extraction test (PBET) (Ruby et al., 1993) to assess the in vitro bioaccessibility (IVBA) of Pb in various of vegetables, Intawongse and Dean (2008) reported gastric IVBA Pb of 7 to 27% and intestinal IVBA Pb of 20 to 61%. Similar results are reported by Fu and Cui (2013), Pan et al. (2016), and Zhuang et al. (2018). Assuming 100% relative bioavailability in risk assessment calculations likely overestimates the contribution of vegetable Pb toward exposure while gardening. More information is needed on vegetable Pb uptake and bioavailability to determine whether this pathway is truly a major concern for gardeners.

## Soil Health Based Assessment and Management of Soil Lead

Risk-based soil screening levels (SSL) are used by many regulatory agencies to assess urban soil Pb. Soil screening levels are very conservative measures, assume worse case scenarios, and are considered "one-way tests." If the soil contaminant is less than the SSL, then no further assessment is needed, but if the soil contaminant exceeds the SSL, further investigation is required. The SSL for soil Pb varies with land use and between regulatory agencies. Soil Pb SSLs are lower for residential than industrial land use because exposure factors are greater in residential settings.

**Table 7.1** Comparison of international residential and industrial lead (Pb) soil screening limits (SSLs).

| | Agency | Residential SSL (mg Pb $kg^{-1}$ soil) | Industrial SSL (mg Pb $kg^{-1}$ soil) | Reference |
|---|---|---|---|---|
| ***Country*** | | | | |
| Australia and New Zealand | NEPC | 300 | 1500 | National Environment Protection Council, 2011 |
| Canada | CCME | 140 | 600 | Canadian Council of Ministers of the Environment, 1999 |
| Hong Kong | EPD | 248 | 2290 | Environmental Protection Department, 2007 |
| UK | EA | 310 | 2300 | Ministry of Infrastructure and the Environment, 2015 |
| USA | EPA | 400 (play areas)<br>1200 (rest of yard) | 800 | United States Environmental Protection Agency, 2019 |
| ***States*** | | | | |
| Michigan | MDEQ | 400 | 900 | Michigan Department of Environmental Quality, 2016 |
| California | OEHHA | 80 | 320 | California Environmental Protection Agency, 2009 |
| Florida | FDEP | 400 | 1400 | University of Florida, 2005 |
| Kansas | KDHE | 400 | 1000 | Kansas Department of Health and Environment, 2015 |

The USEPA Lead Rule (USEPA, 1996) and HUD guidelines (HUD, 2012) suggest a residential Pb hazard exists if bare soil has a Pb concentration of 400 mg Pb $kg^{-1}$ soil in children's play areas and 1200 mg Pb $kg^{-1}$ soil in the remainder of the yard. These values are derived assuming incidental ingestion of soil and dust are the risk drivers. Most states in the U.S. have adopted the USEPA SSL of 400 mg Pb $kg^{-1}$ soil.

Use of urban soil for gardening does not present the same exposure scenario as bare soil in children's play areas. Soil Therefore, in the absence of federal guidance, soil testing laboratories, university extension services, and state health departments have established screening levels for garden soils. In general, those recommendations classify soil Pb levels in three categories: (1) no concern, (2) do not grow leafy green or root crops, and (3) do not even grow fruits (Table 7.2).

**Table 7.2** Recommendations for food and soil management practices to reduce human lead exposure based on soil Pb levels in lead-contaminated soils from federal agencies, land-grant universities or regional colleges.

| | Soil Pb Limit (mg Pb $kg^{-1}$ soil) | | | Food Practices | | Soil Practices | | | | | | | | | |
|---|---|---|---|---|---|---|---|---|---|---|---|---|---|---|---|
| | | | | | | Reducing Ingestion | | | | | | Remediation | | | |
| **Organization** | **No concern level** | **Do not grow leafy greens/roots** | **Do not grow vegetables** | **Peel and remove leaves** | **Wash with cleaning solution** | **Wear gloves/ wash hands** | **Not near roads and/or buildings** | **Raised beds** | **Avoid tracking** | **Cover Bare Soil** | **Limit children's contact** | **Add P** | **Raise pH** | **Dilute with tillage** | **Add OM** |
| UNH Cooperative Extension | 65 | 180 | 450 | x | | | | | | | x | | | | |
| University of Missouri | 50 | 400 | 1200 | | | | | | | x | | | | | |
| Rutgers University | 100 | 100 | 400 | x | x | x | x | x | x | x | | | x | | |
| University of Maryland Extension | 50 | 50 | 400 | x | | x | x | | | | x | x | | | x |
| University of California Cooperative Extension | 80 | – | 80 | | | x | | x | | x | | | x | x | |
| University of Illinois Extension | 100 | 300 | 500 | | x | x | x | | | x | | | | | |

| Organization | Soil Pb Limit (mg Pb $kg^{-1}$ soil) | | | Food Practices | | Soil Practices: Reducing Ingestion | | | | | | Soil Practices: Remediation | | | |
|---|---|---|---|---|---|---|---|---|---|---|---|---|---|---|---|
| | No concern level | Do not grow leafy greens/roots | Do not grow vegetables | Peel and remove leaves | Wash with cleaning solution | Wear gloves/ wash hands | Not near roads and/or buildings | Raised beds | Avoid tracking | Cover Bare Soil | Limit children's contact | Add P | Raise pH | Dilute with tillage | Add OM |
| University of Minnesota Extension | 100 | – | 300 | x | x | | x | | | | | | x | | x |
| Kansas State Extension | – | 200 | 1000 | x | x | x | | x | x | x | | | x | | x |
| Brooklyn College | 100 | 100 | 400 | x | | x | x | x | x | x | x | x | x | | x |
| Connecticut Agricultural Experiment Station | 100 | 100 | 400 | | | | x | | x | x | | | | | |
| Cornell Waste Management Institute | 63 | – | 400 | | | | | | | | | | | | |
| Penn State Extension | 150 | 400 | 1000 | x | x | x | x | x | | x | x | x | x | | x |
| Michigan State University Extension | 300 | 300 | 400 | x | | x | | x | | x | x | | x | | x |
| USEPA TRW | 100 | 100 | 1200 | x | | x | | x | x | x | x | | x | | x |
| USEPA Region 9 | 80 | 500 | 1200 | | | | | | | | | | | | |

Additional food and soil management practices are recommended to reduce Pb exposure for soils with Pb content above the no concern level. Table 7.2 also shows a lack of agreement on soil Pb standards and management practices across states, indicating a more unified interpretation program is needed.

We suggest a soil health-based framework for Pb contamination should be based on human and ecological health. In human health risk assessment, the exposure to contaminated urban soils is quantified by considering the magnitude, frequency, and duration of exposure for the receptors and exposure pathways selected for quantitative evaluation. The main exposure pathway for urban Pb is incidental ingestion associated with hand to mouth activities, ingestion of fugitive dust, and ingestion of soil associated with homegrown produce (S. Brown et al., 2016; Chaney & Ryan, 1994). The following risk-based equation is used to quantify average daily chemical contaminant intake associated with incidental ingestion of soil (USEPA, 1989):

$$\mathrm{CDI} = \left(\mathrm{CS} \times \mathrm{IR} \times \mathrm{CF} \times \mathrm{FI} \times \mathrm{EF} \times \mathrm{ED}\right) / \left(\mathrm{BW} \times \mathrm{AT}\right) \quad (7.1)$$

where

CDI = Chemical contaminant daily intake (mg $kg^{-1}$ $d^{-1}$)
CS = Chemical contaminant concentration in soil (mg Pb $kg^{-1}$ soil)
IR = Soil ingestion rate (mg soil $d^{-1}$)
CF = conversion factor ($10^{-6}$ kg $mg^{-1}$)
FI = Fraction ingestion from contaminated source (unitless)
EF = Exposure frequency (days $yr^{-1}$)
ED = Exposure duration (yr)
BW = Body weight (kg)
AT = Averaging time (period over which exposure is averaged; days)

Most contaminant risk exposure, including Pb, is through the forms of Pb that are biologically available for absorption or "bioavailable" to humans. Bioavailability refers to the portion of the total quantity of a chemical present that is absorbed by a living organism. Total Pb content is not a good predictor of plant uptake (*i.e.*, phytoavailability) or absorption by human from soil ingestion (*i.e.*, oral bioavailability) (McBride et al., 2014; Yan et al., 2017). Bioavailable Pb, the portion of Pb dose that enters systemic circulation from the gastrointestinal tract (*i.e.* absorbed dose), influences exposure and human health risk. Bioavailability can also be expressed in relative terms, the ratio of the absorbed fraction from the exposure medium (i.e., Pb-contaminated soil) to the absorbed fraction from the dosing medium used in the critical toxicity study (*i.e.* lead acetate). For incidental ingestion, the above CDI exposure calculation (Equation 1) can be modified by incorporating relative bioavailability (RBA) (Basta et al., 2001) as shown in Equation 2:

$$CDI = (CS \times IR \times CF \times FI \times EF \times ED \times RBA)/(BW \times AT) \quad (7.2)$$

where
RBA = Relative bioavailability (unitless, 0.0 to 1.0)

Contaminant bioavailability should be the basis of assessing risk from heavy metal. Traditionally soil RBA for Pb has been determined through animal feeding studies, which are time-consuming and expensive. Extensive research during the past two decades has led to development and adoption of in vitro bioaccessibility assays to predict RBA Pb to adjust exposure for human health risk assessment (Drexler & Brattin, 2007; Intawongse & Dean, 2006; Juhasz et al., 2007; Ruby et al., 1993; Scheckel et al., 2009; Schroder et al., 2004). These in vitro methods consist of a gastric and occasionally intestinal phase where simulated digestive fluid is combined with soil and mixed at body temperature to mimic digestion. In these stimulated gastrointestinal conditions, a portion of the total soil Pb dissolves. This in vitro soluble, or bioaccessible, Pb (IVBA Pb) represents the fraction of Pb that is made available for absorption in the gut, which can be used to predict RBA. There are several published in vitro bioaccessibility assays that can be used for risk assessments and research (Table 7.3). In general, data from these studies must be converted into relative bioavailability data for exposure assessment in human health risk assessment. This requires a regression equation where RBA Pb is calculated from IVBA Pb. For the best results, bioaccessibility methods should be calibrated with appropriate animal RBA Pb data and reported with a regression model predicting RBA Pb from IVBA Pb.

Biological, chemical, and physical characteristics are collectively evaluated by soil health assessments. Several of those soil properties, including soil pH, SOM, and clay type and content, increase or decrease metal bioavailability to plant, human, and other ecological receptors (Adriano, 2001; Basta et al., 2005; McLaughlin, 2001; Peijnenburg & Jager, 2003). Soil pH, often termed the master variable, affects a vast array of soil biogeochemical processes including metal bioaccessibility and bioavailability. Soil pH affects mineral dissolution rate and solubility of metal solid phases in soil (Basta et al., 2005). In general, metal mineral dissolution and solubility drive plant uptake. Dissolution and solubility increase as soil pH decreases. Soil organic matter decreases metal availability to plants and a lesser extent, humans by chelation reactions and sorption of metal cations to negatively charged ion exchange sites. Adsorption of Pb to soil clay minerals and cation exchange sites or strong specific adsorption of Pb to oxyhydroxide surfaces of amorphous Fe and Al oxides reduces metal bioavailability to both plants and

**Table 7.3** Published in vivo-in vitro correlation (IVIVC) studies evaluating the ability of in vitro gastrointestinal methods to predict soil Pb bioavailability.

| Method | Animal | Endpoint | Number of Soils | Source of Lead | Gastric IVIVC | Reference | Notes |
|---|---|---|---|---|---|---|---|
| Physiologically Based Extraction Test (PBET) | Sprague-Dawley rat | Blood | 7 | Mining, smelting | RBA† = 1.4(IVBA)‡ – 3.2 | Ruby et al., 1996 | Method not adopted for regulatory use. |
| USEPA Method 1340 | Swine | Blood | 19 | Mining, smelting | RBA = 0.878(IVBA) – 0.028 | USEPA, 2007 | Method adopted by USEPA (2013) provides official guidance for USA. Limitations potentially include the overestimation of RBA values in P-treated soils. |
| USEPA Method 1340 | Swine | Blood | 19 | Mining, smelting | RBA = 1.314(IVBA) −1.607 | Drexler & Brattin, 2007 | Not suitable for P-treated soils. Limitations potentially include crude regression equation. |
| Unified BARGE Method (UBM) | Swine | Bone | 19 | Mining, smelting | RBA= 1.00(IVBA) + 4.75 | Denys et al., 2012 | BARGE 2016, ISO 17924; EU method and international standard. |
| In vitro Gastrointestinal (IVG) Method | Swine | Blood | 18 | Residential, mining, smelting | RBA= 1.22(IVBA) + 12.4 | Schroder et al., 2004 | Method not adopted for regulatory use. |

† RBA = Relative Bioavailability.
‡ IVBA = In vitro Bioaccessibility.

**Table 7.4** Soil health practices to manage soil lead.

| Practice | Pathways addressed | Mechanism |
|---|---|---|
| P fertilization | Human bioavailability | Formation of insoluble P-Pb complexes |
| Maintaining year-round vegetative/mulch cover | Incidental ingestion | Prevents soil splash; reduces dust |
| Reduce tillage | Incidental ingestion | Reduces dust; maintains aggregate stability |
| Organic matter addition | Plant bioavailability; incidental ingestion | Dilution; fresh residues promote aggregation; mature residues improve soil moisture through increased WHC; form complexes with Pb |
| Raise pH to neutral | Plant bioavailability | Reduces Pb solubility |

humans (Bohn et al., 1985; Kabata-Pendias, 2010; Sparks, 2003; Sposito, 2016). Many soil health management practices will reduce human exposure terms in Equation 2. These include increasing vegetative cover to reduce soil dust ingestion, increasing SOM, avoiding strongly acidic pH values, and reducing clay loss via erosion (Table 7.4).

Bioavailability-based remediation does not remove soil contaminants, but rather reduces their bioavailability and human health impacts associated with exposure (Interstate Technology and Regulatory Council, 2017). Excavation and replacement of contaminated soil are expensive and often ecologically destructive, thus making use of soil amendments to reduce bioavailable Pb a cost-effective alternative. Amendments already being used to improve soil physical, chemical and biological properties, such as composts, manures, municipal biosolids, alkaline materials, and phosphate fertilizers, can reduce Pb bioavailability (Brown et al., 2016; Henry et al., 2015; Scheckel et al., 2013). A common mechanism in these treatments is formation of insoluble Pb minerals which do not dissolve in soil solution or the acidic gastrointestinal tract. (Scheckel et al., 2013). Amendment efficacy depends on composition, soil Pb form, and conditions, such as pH. Basta and McGowen (2004) reported that adding soluble P reduced Pb mobility in mining-contaminated soil, but adding triple super phosphate or bone meal had little effect in reducing bioaccessible Pb in calcareous urban soils from Cleveland (Obrycki et al., 2016). The authors theorized that an abundance of basic cations or a high soil pH limited the formation of Pb-phosphate compounds. Analyzing soils with an in vitro bioaccessibility assay before and after treatment could be used to show efficacy of the soil treatments.

## Practical Assessment of Soil Lead

### Sampling

Our recommended sampling scheme is modified from USEPA Superfund and HUD sampling recommendations (HUD, 2012; USEPA, 1996).

1. **Definitions**
   a) Bare soil: soil which is not covered by grass or other plants, mulches, gravel, or similar coverings.
   b) Dripline area: the area within 3 feet of a building perimeter wall.
   c) Play area: areas of frequent soil contact by children < 6 yr of age.

2. **Defining sampling areas**

   **Gardens**: If possible, take one composite sample from each bed. Where numerous beds exist, group them by location (e.g., beds closer and further from roads) or by similar material. If soil inside the growing areas has been amended and is of different composition than the surrounding yard, sample the yard as well.

   **Yards**: Divided and sample in three areas: children's play areas, non-play areas in the former dripline, and the rest of the yard. USEPA and HUD guidelines claim that only bare soil needs to be sampled, however, bare soil areas may change over time, so more widespread sampling is recommended. Children's play areas are sampled separately because increased exposure to soil in these areas means they present a hazard at lower soil concentrations. Soil in children's play areas present a health hazard at 400 mg Pb $kg^{-1}$ soil, while soil in the dripline and remainder of the yard present a hazard at 1200 mg Pb $kg^{-1}$ soil.

   **Post-demolition properties and other contaminated sites**: should be divided into two areas for sampling: those of suspected contamination (such as the former dripline), and the rest of the site. Sampling the dripline separately will allow hotspots to be identified and can target remediation to smaller portions of the site (Figure 7.3).

3. **Equipment, Materials, and Supplies**
   a) Coring Tool: a conventional soil T probe may be used to collect soil cores. Alternative coring equipment include plastic syringes with the ends cut off, spoons, trowels and bulb corers. The tool should be able to take a sample that is at least 1.5 cm wide, uniform in width and depth, and can be easily cleaned or disposed.
   b) Sampling containers for composite samples: Paper or plastic bags, centrifuge tubes, and glass containers are acceptable. Containers should not spill or leak (double-bag) and should be clearly labeled and should be

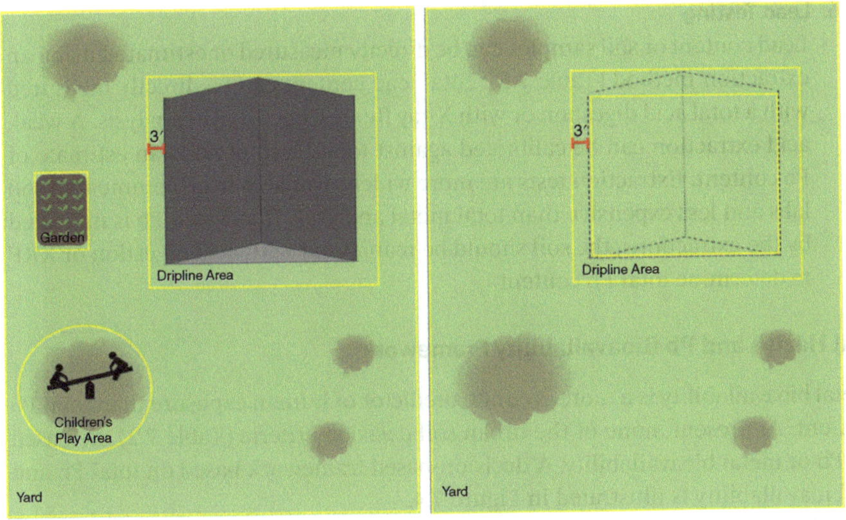


**Figure 7.3** Sampling areas for current residences (left) and post-demolition sites (right).

identified with the following information: Site ID, sampling zone (e.g., dripline, play area, etc.), and depth.

c) Measuring device, flags or stakes, and string to delineate sampling zones

**4. Sampling Scheme**

Soil lead distribution can be very heterogeneous and it is important to get an accurate representation of the entire site while not having excessive costs of numerous samples. For this reason, composite sampling is used.

The number of composite samples recommended depends on the size of the sampling area and resources available. USEPA recommends taking 4 to 6 composite samples consisting of 4 to 6 subsamples from each sampling zone for areas up to 0.25 ha. User discretion should be used when determining the number of composite samples, based on site size and available resources. Subsamples should be taken in a random fashion.

**5. Sampling**

Children and adults are exposed to lead in soil primarily through the immediate surface soil. Lead and other metals are also likely to be concentrated near the surface. Thus, only the top 1.3 to 2.5 cm of soil should be sampled in yards and contaminated sites. Garden soils are typically cultivated and soil Pb content is similar in the surface and subsurface soil; samples should be taken to a depth of 10 to 15 cm.

6. **Lead Testing**
   Lead content of soil samples can be directly measured or estimated using an extraction method (Table 7.5). Total lead content can be directly measured with a total acid digestion or with X-ray fluorescence (XRF) analysis. A weak acid extraction can be calibrated against total Pb to provide an estimate of Pb content. Extraction tests are more widely available from commercial soil labs and less expensive than total metal analyses. If elevated Pb is indicated by the extractions, the soil should be reanalyzed with acid digestion or XRF to determine total Pb content.

## Soil Health and Pb Bioavailability Framework

Metal bioavailability is a more accurate predictor of human exposure than total Pb content. At present, none of the urban soil decision criteria (Table 7.2) are based on Pb or metal bioavailability. A decision-based framework based on total Pb and Pb bioavailability is illustrated in Figure 7.4.

In this approach, total Pb content is compared to risk-based soil screening level (SSL) used by many regulatory agencies to assess urban soil Pb. If the soil contaminant is less than the SSL, then no further assessment is needed. If the soil contaminant exceeds the SSL, then further investigation is required. Screening levels are very conservative measures and assume worse case scenarios. If soil Pb exceeds the screening limit, the exposure equation can be adjusted with soil-specific bioavailability data gathered from in vitro bioaccessibility tests or animal feeding studies. Figure 7.4 provides an approach for evaluating soil Pb risk on residential sites but can be used for other areas if the screening limits and bioavailability assumptions are adjusted. The USEPA default bioavailability for soil Pb is 60% relative bioavailability (RBA), so the residential soil screening limit of 400 mg $kg^{-1}$ total Pb would be equivalent to 240 mg/kg of relative bioavailable Pb. The RBA Pb is then calculated from in vitro bioaccessible Pb measured in the lab with the corresponding in vivo-in vitro correlation in Table 7.3 (e.g., RBA = IVBA $\times$ 0.878 – 0.28 for USEPA Method 1340). If a soil exceeds the screening limit, but bioaccessibility tests reveal bioavailable Pb $<$ 240 mg Pb $kg^{-1}$ soil (*i.e.*, default USEPA level), then further investigation is not needed. If bioavailable Pb $>$ 240 mg Pb $kg^{-1}$ soil, then the soil can be treated to reduce Pb bioavailability (Table 7.6). Lead-contaminated soil treated with amendments should be tested to determine if the amendment reduced bioavailable Pb. Several studies have shown that USEPA Method 1340 potentially over predicts lead RBA in soils amended with P treatments (Obrycki et al., 2016). Consequently, Method 1340 is not an accurate predictor of the reduction of lead RBA achieved by soil amendments. Possible approaches to evaluate amended Pb-contaminated soils include using a modified Method 1340 with extraction solution at pH 2.5 instead of 1.5, or simply using an alternative method with a higher extraction pH (e.g., PBET, OSU IVG). If

**Table 7.5** Comparison of methods used to measure or estimate total Pb.

| Test | Reference | Type | Extraction or Direct Measurement | Approximate cost in 2018 (USD) | Pros | Cons | Best Uses |
|---|---|---|---|---|---|---|---|
| USEPA 3051a | USEPA, 2007 | Total | 1:3 HCl:$HNO_3$ | 35–60 | Standard method | Specialized equipment, cost, time | All, analytic grade data |
| XRF | USDA, 2014 | Total | X-ray fluorescence | 5–20 | Quick, inexpensive | Specialized equipment | Screen, analytic grade data |
| Modified Morgan | Wharton et al., 2012 | Weak acid extraction | Ammonium acetate, pH 4.8 | 0–20 | Can be included with nutrient test | Estimate | Screen, add on to nutrient test |
| Mehlich 3 | Wharton et al., 2012 | Weak acid extraction | Acetic acid and minerals, pH 2.0 | 0–20 | Can be included with nutrient test | Estimate | Screen, add on to nutrient test |
| Nitric Acid | Wharton et al., 2012 | Weak acid extraction | 1 M nitric acid | 15 | Inexpensive | Estimate | Screen |

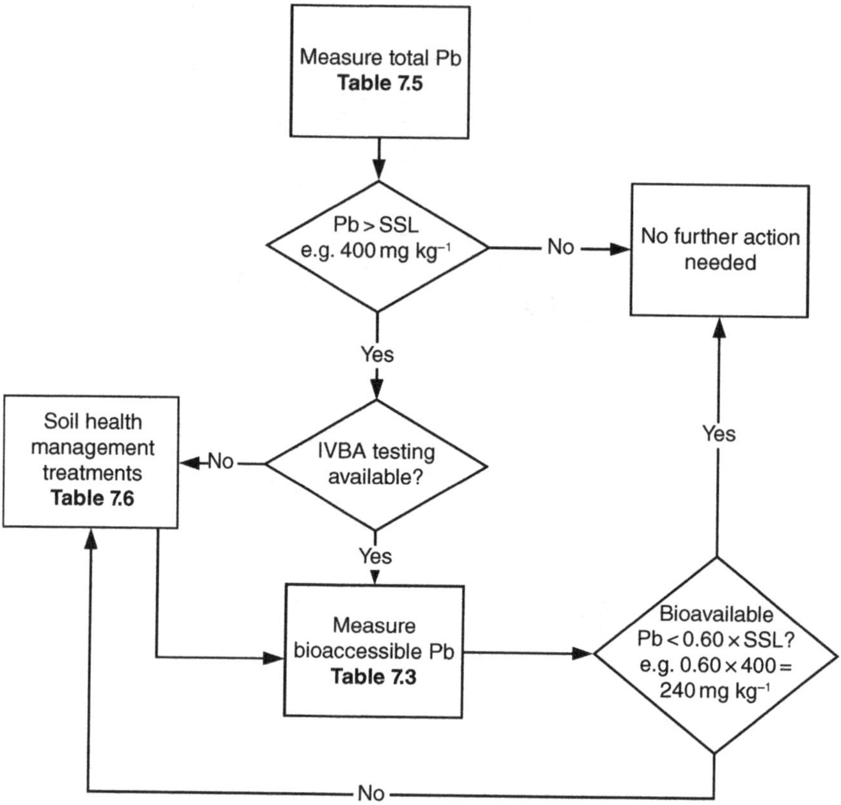


**Figure 7.4** Decision Tree for Evaluating Soil Pb.

USEPA Method 1340 with extraction pH 2.5 is used, then RBA Pb in the amended soil can be calculated from Drexler and Brattin (2007) using the regression equation for pH 2.5 (RBA = 1.314 × IVBA – 1.607) (Table 7.3). Where resources do not exist to conduct bioavailability testing (such as a garden), soil amendments (Table 7.6) can be applied to moderately contaminated soils (Table 7.2) to improve soil health and reduce Pb bioavailability.

## Case Study

Yard soils in the Tar Creek National Priorities List Superfund site, where mean lead concentrations were 623 mg Pb $kg^{-1}$ soil (well above the USEPA residential soil screening limit) were evaluated by Brown et al. (2007). To reduce bioavailability, several soil amendments were evaluated alone and in combination (Table 7.7). In vitro bioaccessible Pb (IVBA Pb) was measured with the physiologically based

**Table 7.6** Soil health practices to manage Pb exposure.

| Practice | Pathways addressed | Materials | Application Rates |
|---|---|---|---|
| P fertilization | Human and plant bioavailability | Triple superphosphate, poultry manure, monoammonium and diammonium phosphate. For acidic soils ($pH < 5$), hydroxyapatite, bone meal, and rock phosphate | 1% by dry mass (Brown et al., 2007; Cai et al., 2017; Scheckel et al., 2013) |
| Maintaining year-round vegetative/mulch cover | Incidental ingestion | Cover crops. Organic mulches: wood chips, straw, compost. Inorganic mulches: plastic, landscape fabric, gravel, rubber | Apply at least 5 cm of organic mulch and reapply as needed. |
| Reduce tillage | Incidental ingestion | N/A | N/A |
| Organic matter addition | Plant bioavailability; incidental ingestion | Biosolids, compost, food scraps, yard waste, manure, wood and paper wastes | 10% by dry mass (Brown et al., 2016) |
| Raise pH to neutral | Plant bioavailability | Ag lime, ash | Provided by commercial soil test |

**Table 7.7** Comparison of in vitro bioaccessible soil lead reductions in response to different amendments.

| Treatment | IVBA Bioaccessible Pb | Estimated Relative Bioavailable Pb | Estimated Bioavailable Pb | Reduction |
|---|---|---|---|---|
| | % | % | mg Pb $kg^{-1}$ soil | % |
| Control | 84 | 114 | 623 (using 100% RBA) | N/A |
| Diammonium phosphate | 55 | 74 | 460 | 26 |
| Biosolids | 52 | 70 | 434 | 30 |
| Compost | 55 | 74 | 460 | 26 |
| Compost + Alum | 36 | 47 | 294 | 53 |

extraction test (PBET) (Ruby et al., 1996). Relative bioavailable Pb was calculated from IVBA Pb data using the regression equation in Table 7.3. Bioavailable Pb in control soils was initially calculated as 114%, or 623 mg Pb $kg^{-1}$ soil (100% of soil Pb). This is greater than the screening limit of 240 mg bioavailable Pb $kg^{-1}$ soil (Figure 7.4). Treatment with phosphates and organic materials reduced Pb bioavailability. The most effective treatment was diammonium phosphate (DAP) and compost, which reduced bioavailable Pb by 53%. By taking advantage of multiple mechanisms, combining amendments can maximize both bioavailability reductions and soil health benefits.

## Summary and Conclusions

To date, most soil health assessments have been associated with crop productivity, but they can also be very useful for many other land uses including remediation and for connecting soil and human health. This chapter illustrates health and remediation issues with a focus on lead, one of the most common urban legacy soil contaminants due to mining, refining and industrial processes, and ubiquitous inclusion in gasoline and paint products throughout much of the 20th Century. Soil assessment of Pb is currently conducted by evaluating exposure in human health risk assessment. Risk-based soil screening levels (SSL) based on human exposure are used by many federal and state regulatory agencies, but they are very conservative and assume worst-case scenarios including conservative default human exposure values. This chapter provides a novel risk-based soil health approach to management of soil lead including exposure pathways, risk assessment, and restoration strategies to improve soil health and reduce human health exposure and risk. A case study is used to illustrate a potential soil health framework for contaminants based on human and ecological health. The risk-based framework is applied to evaluate the use of soil amendments to remediate Pb contaminated soils.

## References


Adriano, D. C. (2001). Ecological and health risks of metals. In *Trace elements in terrestrial environments* (p. 133–165). Springer, New York.

Agbenin, J. O., Danko, M., & Welp, G. (2009). Soil and vegetable compositional relationships of eight potentially toxic metals in urban garden fields from northern Nigeria. *Journal of the Science of Food & Agriculture*, 89(1), 49–54.

Alexander, P. D., Alloway, B. J., & Dourado, A. M. (2006). Genotypic variations in the accumulation of Cd, Cu, Pb and Zn exhibited by six commonly grown vegetables. *Environmental Pollution* 144(3), 736–745. doi:10.1016/j.envpol.2006.03.001

Alloway, B. J., Thornton, I., Smart, G. A., Sherlock, J. C., & Quinn, M. J. (1988). Metal availability. *Science of the Total Environment*, 75(1), 41–69.

Barltrop, D., & Khoo, H. E. (1975). The influence of nutritional factors on lead absorption. *Postgraduate Medical Journal*, 51(601), 795–800. doi:10.1136/pgmj.51.601.795

Basta, N., & McGowen, S. (2004). Evaluation of chemical immobilization treatments for reducing heavy metal transport in a smelter-contaminated soil. *Environmental Pollution*, 127(1), 73–82. doi:10.1016/S0269-7491(03)00250-1

Basta, N., Ryan, J. A., & Chaney, R. L. (2005). Trace element chemistry in residual-treated soil: Key concepts and metal bioavailability. *Journal of Environmental Quality*, 34(1), 49–63. doi:10.2134/jeq2005.0049dup

Basta, N. T., Rodriguez, R. R., & Casteel, S. W. (2001). Bioavailability and risk of arsenic exposure by the soil ingestion pathway. In W. T. Frankenberger (Ed.) *Environmental chemistry of arsenic* (pp. 137–160). New York: Marcel Dekker, Inc.

Bohn, H. L., McNeal, B. L., & O'Connor, G. A. (1985). *Soil chemistry*. New York: Wiley.

Brown, M. J., & Margolis, S. (2012). Lead in drinking water and human blood lead levels in the United States. Morbidity and Mortality Weekly Report. August 10, 2012. Atlanta, GA: Centers for Disease Control and Prevention.

Brown, S. L., Chaney, R. L., & Hettiarachchi, G. M. (2016). Lead in urban soils: A real or perceived concern for urban agriculture? *Journal of Environmental Quality*, 45(1), 26–36. doi:10.2134/jeq2015.07.0376

Brown, S. L., Compton, H., & Basta, N. (2007). Field test of in situ soil amendments at the tar creek national priorities list superfund site. *Journal of Environmental Quality*, 36(6), 1627–1634. doi:10.2134/jeq2007.0018

Cai, M. F., McBride, M. B., Li, K. M., & Li, Z. (2017). Bioaccessibility of As and Pb in orchard and urban soils amended with phosphate, Fe oxide and organic matter. *Chemosphere*, 173, 153–159. doi:10.1016/j.chemosphere.2017.01.049

California Environmental Protection Agency. (2009). Revised California Human Health Screening Levels for Lead and Beryllium. Sacramento: California Environmental Protection Agency. https://oehha.ca.gov/risk/soils091709

Canadian Council of Ministers of the Environment. (1999). Canadian soil quality guidelines for the protection of environmental and human health. Ottowa: Canadian Council of Ministers of the Environment. http://ceqg-rcqe.ccme.ca/download/en/269

Chaney, R. L., & Ryan, J. A. (1994). Risk based standards for arsenic, lead and cadmium in urban soils: Summary of information and methods developed to estimate standards for Cd, Pb and As in urban soils. Frankfurt: Society for Chemical Engineering and Biotechnology.

Clark, J. J., & Knudsen, A. C. (2013). Extent, characterization, and sources of soil lead contamination in small-urban residential neighborhoods. *Journal of Environmental Quality*, 42(5), 1498–1506. doi:10.2134/jeq2013.03.0100

Cuyahoga County Board of Health. (2017). *City of Cleveland neighborhoods confirmed elevated blood lead levels (EBLs) for children < 6 years of age, January 12017-December 13, 2017*. Cleveland, OH. Retrieved from https://www.ccbh.net/wp-content/uploads/2019/01/Report-2017-Children-residing-in-the-City-of-Cleveland.pdf

Datko-Williams, L., Wilkie, A., & Richmond-Bryant, J. (2014). Analysis of U.S. soil lead (Pb) studies from 1970 to 2012. *Science of the Total Environment*, 468, 854–863. doi:10.1016/j.scitotenv.2013.08.089

Deluca, J., Hardy, C. A., Burright, R. G., Donovick, P. J., & Tuggy, R. L. (1982). The effects of dietary fat and lead ingestion on blood lead levels in mice. *Journal of Toxicology & Environmental Health*, 10(3), 441–447. doi:10.1080/15287398209530266

Denys, S., Caboche, J., Tack, K., Rychen, G., Wragg, J., Cave, M., & Feidt, C. (2012). In vivo validation of the unified BARGE method to assess the bioaccessibility of arsenic, antimony, cadmium, and lead in soils. *Environmental Science & Technology*, 46(11), 6252–6260. https://doi.org/10.1021/es3006942. doi:10.1021/es3006942

Doelman, P., & Haanstra, L. (1979a). Effect of lead on soil respiration and dehydrogenase activity. *Soil Biology & Biochemistry*, 11(5), 475–479. doi:10.1016/0038-0717(79)90005-1

Doelman, P., & Haanstra, L. (1979b). Effects of lead on the soil bacterial microflora. *Soil Biology & Biochemistry*, 11(5), 487–491. doi:10.1016/0038-0717(79)90007-5

Doelman, P., & Haanstra, L. (1984). Short-term and long-term effects of cadmium, chromium, copper, nickel, lead and zinc on soil microbial respiration in relation to abiotic soil factors. *Plant Soil*, 79(3), 317–327. doi:10.1007/BF02184325

Drexler, J. W., & Brattin, W. J. (2007). An in vitro procedure for estimation of lead relative bioavailability: With validation. *Human Ecology Risk Assessment*, 13(2), 383–401. doi:10.1080/10807030701226350

Duzgoren-Aydin, N. S., Wong, C. S. C., Aydin, A., Song, Z., You, M., & Li, X. D. (2006). Heavy metal contamination and distribution in the urban environment of Guangzhou, SE China. *Environmental Geochemistry & Health*, 28(4), 375–391. doi:10.1007/s10653-005-9036-7

Environmental Protection Department. (2007). Guidance manual for use of risk-based remediation goals for contaminated land management. Hong Kong: Environmental Protection Department. https://www.epd.gov.hk/epd/sites/default/files/epd/english/environmentinhk/waste/guide_ref/files/gme.pdf

Fan, Y., Li, H., Xue, Z. J., Zhang, Q., & Cheng, F. Q. (2017). Accumulation characteristics and potential risk of heavy metals in soil-vegetable system under greenhouse cultivation condition in Northern China. *Ecological Engineering*, 102, 367–373.

Farrell, K. P., Brophy, M. C., Chisolm, J. J., Rohde, C. A., & Strauss, W. K. (1998). Soil lead abatement and children's blood lead levels in an urban setting. *American Journal of Public Health*, 88(12), 1837–1839. doi:10.2105/AJPH.88.12.1837

Finster, M. E., Gray, K. A., & Binns, H. J. (2004). Lead levels of edibles grown in contaminated residential soils: A field survey. *Science of the Total Environment*, 320(2-3), 245–257. doi:10.1016/j.scitotenv.2003.08.009

Fu, J., & Cui, Y. S. (2013). In vitro digestion/Caco-2 cell model to estimate cadmium and lead bioaccessibility/bioavailability in two vegetables: The influence of cooking and additives. *Food Chemistry & Toxicology*, 59, 215–221. doi:10.1016/j.fct.2013.06.014

Gao, M., Song, W., Zhou, Q., Ma, X., & Chen, X. (2013). Interactive effect of oxytetracycline and lead on soil enzymatic activity and microbial biomass. *Environmental Toxicology and Pharmacology*, 36(2), 667–674.

Hauptman, M., Bruccoleri, R., & Woolf, A. D. (2017). An update on childhood lead poisoning. *Clinical Pediatric Emergency Medicine*, 18(3), 181–192. doi:10.1016/j.cpem.2017.07.010

Henry, H., Naujokas, M. F., Attanayake, C., Basta, N., Cheng, Z. Q, Hettiarachchi, G. M., & Scheckel, K. G. (2015). Bioavailability-based in situ remediation to meet future lead (Pb) standards in urban soils and gardens. *Environmental Science & Technology*, 49(15), 8948–8958. doi:10.1021/acs.est.5b01693

Hough, R. L., Breward, N., Young, S. D., Crout, N. M. J., Tye, A. M., Moir, A. M., & Thornton, I. (2004). Assessing potential risk of heavy metal exposure from consumption of home-produced vegetables by urban populations. *Environmental Health Perspectives*, 112(2), 215–221.

Housing and Urban Development (HUD). (2012). Guidelines for the evaluation and control of lead-based paint hazards in housing, Second ed. U.S. Department of Housing and Urban Development, Office of Lead-Based Paint Abatement and Poisoning Prevention. Retrieved from http://books.google.com/books?id=7asa09XO1HYC

Intawongse, M., & Dean, J. R. (2006). In-vitro testing for assessing oral bioaccessibility of trace metals in soil and food samples. *TrAC Trends in Analytical Chemistry*, 25(9), 876–886. doi:10.1016/j.trac.2006.03.010

Intawongse, M., & Dean, J. R. (2008). Use of the physiologically-based extraction test to assess the oral bioaccessibility of metals in vegetable plants grown in contaminated soil. *Environmental Pollution*, 152(1), 60–72. doi:10.1016/j.envpol.2007.05.022

Interstate Technology and Regulatory Council. (2017). Bioavailability of contaminants in soil: Considerations for Human Health Risk Assessment (BCS-1). Washington, DC: Interstate Technology and Regulatory Council. Retrieved from https://bcs-1.itrcweb.org

James, H. M., Hilburn, M. E., & Blair, J. A. (1985). Effects of meals and meal times on uptake of lead from the gastrointestinal-tract in humans. *Human Toxicology*, 4(4), 401–407.

Johnson, C.C., Demetriades, A., Locutura, J., & Ottesen, R. T. (2012). Mapping the chemical environment of urban areas. New York: John Wiley & Sons.

Juhasz, A. L., Smith, E., Weber, J., Rees, M., Rofe, A., Kuchel, T., & Naidu, R. (2007). Comparison of in vivo and in vitro methodologies for the assessment of arsenic bioavailability in contaminated soils. *Chemosphere*, 69(6):961–966. doi:10.1016/j.chemosphere.2007.05.018

Kabata-Pendias, A. (2010). Trace elements in soils and plants. Boca Raton: CRC Press. doi:10.1201/b10158

Kansas Department of Health and Environment. (2015). Kansas Department of Health and Environment. Retrieved from https://www.kdheks.gov/remedial/rsk_manual_page.html

Kansas Department of Health and Environment. (2019). National Zinc Company (Cherryvale Zinc Division) Site. Retrieved April 14, 2020. https://www.kdheks.gov/remedial/site_remediation/national_zinc.html

Kelly, J., Thornton, I., & Simpson, P. R. (1996). Urban geochemistry: A study of the influence of anthropogenic activity on the heavy metal content of soils in traditionally industrial and nonindustrial areas of Britain. *Applied Geochemistry*, 11(1-2), 363–370. doi:10.1016/0883-2927(95)00084-4

Khan, A., Khan, S., Khan, M. A., Qamar, Z., & Waqas, M. (2015). The uptake and bioaccumulation of heavy metals by food plants, their effects on plants nutrients, and associated health risk: A review. *Environmental Science Pollution Research International*, 22(18), 13772–13799.

Konopka, A., Zakharova, T., Bischoff, M., Oliver, L., Nakatsu, C., & Turco, R. (1999). Microbial biomass and activity in lead-contaminated soil. *Applied Environmental Microbiology*, 65(5), 2256–2259. doi:10.1128/AEM.65.5.2256-2259.1999

Li, H. B., Li, M. Y., Zhao, D., Zhu, Y. G., Li, J., Juhasz, A. L., & Ma, L. Q. (2018). Food influence on lead relative bioavailability in contaminated soils: Mechanisms and health implications. *Journal of Hazardous Materials.* 358, 427–433. doi:10.1016/j.jhazmat.2018.06.034

Li, J., Huang, Y., Hu, Y., Jin, S., Bao, Q., Wang, F., & Xie, H. (2016). Lead toxicity thresholds in 17 Chinese soils based on substrate-induced nitrification assay. *Journal of Environmental Sciences*, 44, 131–140.

Liao, M., Chen, C. -L., Zeng, L. -S., & Huang, C.-Y. (2007). Influence of lead acetate on soil microbial biomass and community structure in two different soils with the growth of Chinese cabbage (*Brassica chinensis*). *Chemosphere*, 66(7), 1197–1205. doi:10.1016/j.chemosphere.2006.07.046

Liu, Y. J., Naidu, R., & Semple, K. (2017). Measurement of soil lead bioavailability and influence of soil types and properties: A review. *Chemosphere*, 184, 27–42. doi:10.1016/j.chemosphere.2017.05.143

Maddaloni, M., Lolacono, N., Manton, W., Blum, C., Drexler, J., & Graziano, J. (1998). Bioavailability of soilborne lead in adults, by stable isotope dilution. *Environmental Health Perspectives*, 106, 1589–1594.

Marzadori, C., Ciavatta, C., Montecchio, D., & Gessa, C. (1996). Effects of lead pollution on different soil enzyme activities. *Biology and Fertility of Soils*, 22(1-2), 53–58.

McBride, M. B., Shayler, H. A., Spliethoff, H. M., Mitchell, R. G., Marquez-Bravo, L. G., Ferenz, G. S., & Bachman, S. (2014). Concentrations of lead, cadmium and barium in urban garden-grown vegetables: The impact of soil variables. *Environmental Pollution*, 194, 254–261. doi:10.1016/j.envpol.2014.07.036

McLaughlin, M. J. (2001). Bioavailability of metals to terrestrial plants. In Bioavailability of metals in terrestrial ecosystems: Importance of partitioning for bioavailability to invertebrates, microbes, and plants (p. 39–68). Boca Raton: Society of Environmental Toxicology and Chemistry.

Michigan Department of Environmental Quality. (2016). Cleanup criteria and screening levels development and application: Remediation and redevelopment division resource materials. Michigan Department of Environmental Quality, Remediation and Redevelopment Division.

Michigan Department of Health & Human Services. (2018). 2016 Data report on childhood lead testing and elevated blood lead levels: Michigan. Michigan Department of Health & Human Services, Childhood Lead Poisoning Prevention Program (CLPPP). Retrieved from https://www.michigan.gov/documents/lead/2016_CLPPP_Annual_Report_5-1-18_621989_7.pdf

Mielke, H. (1999). Lead in the inner cities. *American Science*, 87(1), 62–73. doi:10.1511/1999.1.62

Mielke, H. W., Adams, J. E., Huff, B., Pepersack, J., Reagan, P., Stoppel, D., & Mielke, P., Jr. (1992). Dust control as a means of reducing inner-city childhood Pb exposure. *Trace Substances in Environmental Health*, 25, 121–128.

Mielke, H. W., Gonzales, C. R., Powell, E. T., Laidlaw, M. A. S., Berry, K. J., Mielke, P. W., & Egendorf, S. P. (2019). The concurrent decline of soil lead and children's blood lead in New Orleans. *Proceedings of National Academy of Science USA*, 116(44), 22058–22064. doi:10.1073/pnas.1906092116

Mielke, H. W., Gonzales, C. R., Smith, M. K., & Mielke, P. W. (1999). The urban environment and children's health: Soils as an integrator of lead, zinc, and cadmium in New Orleans, Louisiana, USA. *Environmental Research*, 81(2), 117–129.

Mielke, H. W., Laidlaw, M. A. S., & Gonzales, C. R. (2011). Estimation of leaded (Pb) gasoline's continuing material and health impacts on 90 US urbanized areas. *Environmental International*, 37(1), 248–257. doi:10.1016/j.envint.2010.08.006

Minca, K. K., & Basta, N. T. (2013). Comparison of plant nutrient and environmental soil tests to predict Pb in urban soils. *Science of the Total Environment*, 445-446, 57–63. doi:10.1016/j.scitotenv.2012.12.008

Minca, K. K., Basta, N., & Scheckel, K. G. (2013). Using the Mehlich-3 soil test as an inexpensive screening tool to estimate total and bioaccessible lead in urban soils. *Journal of Environmental Quality*, 42(5), 1518–1526. doi:10.2134/jeq2012.0450

Ministry of Infrastructure and the Environment. (2015). *Into Dutch soils*. The Netherlands: Ministry of Infrastructure and the Environment. https://rwsenvironment.eu/publish/pages/126603/into_dutch_soils.pdf

Murata, T., Kanao-Koshikawa, M., & Takamatsu, T. (2005). Effects of Pb, Cu, Sb, In and Ag contamination on the proliferation of soil bacterial colonies, soil dehydrogenase activity, and phospholipid fatty acid profiles of soil microbial communities. *Water, Air, and Soil Pollution*, 164(1-4), 103–118.

Mylroie, A. A., Moore, L., Olyai, B., & Anderson, M. (1978). Increased susceptibility to lead toxicity in rats fed semipurified diets. *Environmental Research*, 15(1), 57–64. doi:10.1016/0013-9351(78)90078-6

National Environment Protection Council. (2011). Guideline on investigation levels for soil and groundwater. Queensland, Australia: National Environment Protection Council. http://www.nepc.gov.au/system/files/resources/93ae0e77-e697-e494-656f-afaaf9fb4277/files/schedule-b1-guideline-investigation-levels-soil-and-groundwater-sep10.pdf

Obrycki, J. F., Basta, N., Scheckel, K., Stevens, B. N., & Minca, K. K. (2016). Phosphorus amendment efficacy for in situ remediation of soil lead depends on the bioaccessible method. *Journal of Environmental Quality*, 45(1), 37–44. doi:10.2134/jeq2015.05.0244

Olowoyo, J. O., & Lion, G. N. (2016). Urban farming as a possible source of trace metals in human diets. *South African Journal of Science*, 112(1-2), 30–35.

Pan, W., Kang, Y., Li, N., Zeng, L., Zhang, Q., Wu, J., & Guo, X. (2016). Bioaccessibility of heavy metals in vegetables and its association with the physicochemical characteristics. *Environmental Science Pollution Research International* 23(6), 5335–5341. doi:10.1007/s11356-015-5726-6

Peijnenburg, W. J. G. M., & Jager, T. (2003). Monitoring approaches to assess bioaccessibility and bioavailability of metals: Matrix issues. *Ecotoxicology Environmental Safety*, 56(1), 63–77. doi:10.1016/S0147-6513(03)00051-4

Preer, J. R., Akintoye, J. O., & Martin, M. L. (1984). Metals in downtown Washington, DC gardens. *Biology Trace Elemental Research*, 6(1), 79–91. doi:10.1007/BF02918323

Ruby, M. V., Davis, A., Link, T. E., Schoof, R., Chaney, R. L., Freeman, G. B., & Bergstrom, P. (1993). Development of an in-vitro screening-test to evaluate the in-vivo bioaccessibility of ingested mine-waste lead. *Environmental Science & Technology*, 27(13), 2870–2877. doi:10.1021/es00049a030

Ruby, M. V., Davis, A., Schoof, R., Eberle, S., & Sellstone, C. M. (1996). Estimation of lead and arsenic bioavailability using a physiologically based extraction test. *Environmental Science & Technology*, 30(2), 422–430. doi:10.1021/es950057z

Scheckel, K. G., Chaney, R. L., Basta, N., & Ryan, J. A. (2009). Advances in assessing bioavailability of metal(loid)s in contaminated soils. *Advances in Agronomy*, 104, 1–52.

Scheckel, K. G., Diamond, G. L., Burgess, M. F., Klotzbach, J. M., Maddaloni, M., Miller, B. W., & Serda, S. M. (2013). Amending soils with phosphate as means to mitigate soil lead hazard: A critical review of the state of the science. *Journal of Toxicology Environmental Health B Critical Reviews*, 16(6), 337–380. doi:10.1080/10937404.2013.825216

Schroder, J. L., Basta, N., Casteel, S. W., Evans, T. J., Payton, M. E., & Si, J. (2004). Validation of the in vitro gastrointestinal (IVG) method to estimate relative bioavailable lead in contaminated soils. *Journal of Environmental Quality*, 33(2), 513–521.

Shacklette, H. T., & Boerngen, J. G. (1984). Element concentrations in soils and other surficial materials of the conterminous United States. Paper 1270. Washington, DC: USGS. doi:10.3133/pp1270

Sharma, K., Basta, N., & Grewal, P. S. (2015). Soil heavy metal contamination in residential neighborhoods in post-industrial cities and its potential human exposure risk. *Urban Ecosystems*, 18(1), 115–132. doi:10.1007/s11252-014-0395-7

Shi, W., Becker, J., Bischoff, M., Turco, R., & Konopka, A. (2002a). Association of microbial community composition and activity with lead, chromium, and hydrocarbon contamination. *Applied Environmental Microbiology*, 68(8), 3859–3866. doi:10.1128/AEM.68.8.3859-3866.2002

Shi, W., Bischoff, M., Turco, R., & Konopka, A. (2002b). Long-term effects of chromium and lead upon the activity of soil microbial communities. *Applied Soil Ecology* 21, 169–177. doi:10.1016/S0929-1393(02)00062-8

Smith, D. B., Cannon, W. F., Woodruff, L. G., Solano, F., Kilburn, J. E., & Fey, D. L. (2013). Geochemical and mineralogical data for soils of the conterminous United States. Data series 801. Washington, DC: USGS. doi:10.3133/ds801

Sparks, D. L. (2003). Environmental soil chemistry. Amsterdam; Boston: Academic Press. doi:10.1016/B978-012656446-4/50001-3

Spittler, T. M., & Feder, W. A. (1979). A study of soil contamination and plant lead uptake in Boston urban gardens. *Communications in Soil Science Plant Analysis.* 10, 1195–1210. doi:10.1080/00103627909366973

Spliethoff, H. M., Mitchell, R. G., Shayler, H., Marquez-Bravo, L. G., Russell-Anelli, J., Ferenz, G., & McBride, M. (2016). Estimated lead (Pb) exposures for a population of urban community gardeners. *Environmental Geochemical Health*, 38, 955–971. doi:10.1007/s10653-016-9790-8

Sposito, G. (2016). The chemistry of soils. Oxford: Oxford Univ. Press.

Suhadolc, M., Schroll, R., Gattinger, A., Schloter, M., Munch, J. C., & Lestan, D. (2004). Effects of modified Pb-, Zn-, and Cd- availability on the microbial communities and on the degradation of isoproturon in a heavy metal contaminated soil. *Soil Biology & Biochemistry*, 36, 1943–1954.

United States Department of Agriculture (USDA). (2014). Soil survey field and laboratory methods manual. In R. Burt (Ed)., Soil Survey Investigations Report No. 51, Version 2. Washington, DC: USDA NRCS. Retrieved from https://www.nrcs.usda.gov/Internet/FSE_DOCUMENTS/stelprdb1244466.pdf.

United States Environmental Protection Agency (USEPA). (1989). Risk Assessment Guidance for Superfund (RAGS), Volume I, Human Health Evaluation Manual (Part A). Washington, DC: Office of Emergency and Remedial Response, EPA/540/1-89/002. Retrieved from https://www.epa.gov/risk/risk-assessment-guidance-superfund-rags-part.

United States Environmental Protection Agency (USEPA). (1994). Soil-lead and demographics remedial investigation studies: California Gluch Superfund Site. Washington, DC: USEPA. Retrieved from https://semspub.epa.gov/work/08/318867.pdf.

United States Environmental Protection Agency (USEPA). (1996). *Soil screening guidance: User's guide. Office of Solid Waste and Emergency Response.* Washington, DC: National Service Center for Environmental Publications. Retrieved from https://rais.ornl.gov/documents/SSG_nonrad_user.pdf.

United States Environmental Protection Agency (USEPA). (1999). Common Contaminants Found at Superfund Sites (OSWER 6203.1-17A). Washington, DC: USEPA. Retrieved from https://nepis.epa.gov/Exe/ZyPDF.cgi/10002AYA.PDF?Dockey=10002AYA.PDF.

United States Environmental Protection Agency (USEPA). (2002). Overview of the IEUBK Model for Lead in Children. Washington, DC: USEPA. Retrieved from https://semspub.epa.gov/work/HQ/174574.pdf.

United States Environmental Protection Agency (USEPA). (2007a). Estimation of relative bioavailability of lead in soil and soil-like materials using in vivo and in vitro methods (OSWER 9285.7-77). Washington, DC: National Service Center for Environmental Publications. Retrieved from https://nepis.epa.gov/Exe/ZyPDF.cgi/93001C2U.PDF?Dockey=93001C2U.PDF.

United States Environmental Protection Agency (USEPA). (2007b). Method 3051a. Microwave assisted acid digestion of sediments, sludges, soils, and oils. In: SW-846. Washington, DC: National Service Center for Environmental Publications. Retrievedfromhttps://www.epa.gov/hw-sw846/sw-846-test-method-3051a-microwave-assisted-acid-digestion-sediments-sludges-soils-and-oils.

United States Environmental Protection Agency (USEPA). (2013). Method 1340: In vitro bioaccessibility assay for lead in soil. Washington, DC: USEPA. Retrieved from https://www.epa.gov/hw-sw846/sw-846-test-method-1340-vitro-bioaccessibility-assay-lead-soil.

United States Environmental Protection Agency (USEPA). (2014). Technical Review Workgroup Recommendations Regarding Gardening and Reducing Exposure to Lead-Contaminated Soils. Washington, DC: National Service Center for Environmental Publications. Retrieved from https://nepis.epa.gov/Exe/ZyPDF.cgi/P100JJS3.PDF?Dockey=P100JJS3.PDF.

United States Environmental Protection Agency (USEPA). (2016). West Calumet Housing Complex– East Chicago, Ind. Washington, DC: USEPA. Retrieved from https://www.epa.gov/uss-lead-superfund-site/west-calumet-housing-complex-east-chicago-ind.

United States Environmental Protection Agency (USEPA). (2019). Regional Screening Level (RSL) Composite Worker Soil Table (TR = 1E-06, HQ = 1). Washington, DC: USEPA. Retrieved from https://semspub.epa.gov/work/HQ/199634.pdf.

University of Florida. (2005). Technical Report: Development of Cleanup Target Levels (CTLs) for Chapter 62-777, FAC. Florida Department of Environmental Protection. Retrieved from https://floridadep.gov/waste/district-business-support/documents/technical-report-development-cleanup-target-levels-ctls.

USA Today. (2012). Ghost factories: Poison in the ground. Retrieved from http://usatoday30.usatoday.com/news/nation/lead-poisoning.

Weitzman, M., Aschengrau, A., Bellinger, D., Jones, R., Hamlin, J. S., & Beiser, A. (1993). Lead-contaminated soil abatement and urban childrens blood lead levels. *JAMA*, 269(13), 1647–1654. doi:10.1001/jama.1993.03500130061033

Wharton, S. E., Shayler, H. A., Spliethoff, H. M., Marquez-Bravo, L. G., Ribaudo, L., & McBride, M. B. (2012). A comparison of screening tests for soil Pb. *Soil Science*, 177(11), 650–654. doi:10.1097/SS.0b013e318277718b

Wise, A., & Gilburt, D. J. (1980). The variability of dietary fibre in laboratory animal diets and its relevance to the control of experimental conditions. *Food and Cosmetics Toxicology*, 18(6), 643–648.

World Health Organization/Food and Agriculture Organization of the United Nations (WHO/FAO). (2015). General standard for contaminants and toxins in food and feed (Codex STAN 193-1995). FAO, Rome.

Yang, U. J., Yoon, S. R., Chung, J. H., Kim, Y. J., Park, K. H., Park, T. S., & Shim, S. M. (2012). Water spinach (*Ipomoea aquatic* Forsk.) reduced the absorption of heavy metals in an in vitro bio-mimicking model system. *Food Chemical Toxicology*, 50(10), 3862–3866. doi:10.1016/j.fct.2012.07.020

Zeng, L. -S., Liao, M., Chen, C. -L., & Huang C. -Y. (2006). Effects of lead contamination on soil microbial activity and rice physiological indices in soil–Pb–rice (*Oryza sativa* L.) system. *Chemosphere*, 65(4), 567–574. doi:10.1016/j.chemosphere.2006.02.039

Zhuang, P., Sun, S., Li, Y., Li, F., Zou, B., Li, Y., & Li, Z. (2018). Oral bioaccessibility and exposure risk of metal(loid)s in local residents near a mining-impacted area, Hunan, China. *International Journal of Environmental Research in Public Health*, 15(8), 1573. doi:10.3390/ijerph15081573

# 8

# The Future of Soil Health Assessments: Tools and Strategies

*Kristen S. Veum, Marcio R. Nunes, and Ken A. Sudduth*

## Introduction

Soil health represents the nexus of multiple ecosystem services provided by soil and encompasses important soil functions, including environmental protection and crop productivity (Doran and Zeiss 2000; Stine and Weil, 2002). Basic soil health research focuses on fundamental ecological principles that elucidate the mechanisms underlying soil processes, whereas applied soil health research is designed to address the needs of "on-farm" soil health assessment and interpretation for management decisions. Although both areas of research conceptually overlap, they reflect research needs that are often divergent with respect to goals and objectives, and thus may require different tools and strategies. This chapter will explore novel tools and strategies that are advancing the science of soil health assessment and identify key areas for future research, with an emphasis on producer assessments and agroecosystem management.

In an effort to capture the broad categories of soil function, quantification of soil health typically involves measurement of multiple dynamic chemical, physical, and biological soil properties, referred to as soil health indicators (Karlen et al., 1997; Moebius-Clune et al., 2016). In principle, indicators are paired with knowledge of inherent soil characteristics that define the potential for a given soil and facilitate interpretation (Karlen et al., 1997). An ideal soil health indicator for on-farm, producer assessments must be dynamic, easy to measure,

*Soil Health Series: Volume 1 Approaches to Soil Health Analysis*, First Edition.
Edited by Douglas L. Karlen, Diane E. Stott, and Maysoon M. Mikha.

applicable to field conditions, accessible to many users, and capture a wide range of soil functions (Doran and Parkin 1996; Nortcliff 2002; Andrews et al., 2004). Several chemical, physical, and biological soil properties have been proposed as soil health indicators (Bünemann et al., 2018; Stott 2019), and reliable measurement of these indicators would be beneficial for sustainable agricultural management. However, soils are known for their spatial and temporal variability, particularly with regard to soil health indicators (Jung et al., 2006; Piotrowska-Długosz et al., 2019).

Capturing spatial or temporal variability using traditional laboratory measurements often involves costly and labor-intensive field collection, processing, and/or laboratory analysis, which prohibits the production of high quality, high resolution, field-scale information. First, the act of collecting and processing a soil sample can dramatically alter soil properties and skew results (*e.g.*, Franzluebbers et al., 1996; DeForest 2009; Veum et al., 2019). In addition, traditional laboratory methods may fail to capture real-world processes in a meaningful way, given the controlled, and sometimes extreme, conditions (*e.g.*, temperature, pH, and oxygen) under which they are measured that do not represent environmental conditions in the field. Further, laboratory techniques may require sophisticated instrumentation, advanced technical skills, or expensive consumables that are not readily available. Therefore, gathering sufficient data to accurately quantify and interpret variability in the soil system using traditional methods can be costly, time consuming, and potentially uninformative.

Efforts being made to improve cost-effectiveness in the laboratory include: (1) scaling down methods to: (i) reduce the amount of soil and consumables required, (ii) lower cost, and (iii) increase throughput (*e.g.*, Buyer and Sasser 2012; Giacometti et al., 2014; Hurisso et al., 2018) as well as (2) using low cost instrumentation (*e.g.*, Weil et al., 2003; Bakhshandeh et al., 2019; Franzluebbers and Veum 2020), (3) decreasing reaction times (*e.g.*, Haney et al., 2001; Schomberg et al., 2009), and (4) developing combined assays or panels (*e.g.*, Acosta-Martinez et al., 2018). Cheaper and faster alternatives to bench chemistry methods used in the laboratory include: (i) using spectral reflectance techniques (*e.g.*, Janik et al., 1998; Viscarra Rossel et al., 2016; Johnson et al., 2019) or (ii) pedotransfer functions that generally use low cost, easy to measure properties (*e.g.*, Pachepsky and Van Genuchten 2011; Tóth et al., 2015) to estimate other soil properties that are too costly or time-consuming to measure.

In contrast to more traditional laboratory approaches for assessing soil health indicators, on-the-go sensor technology has the potential to provide high-resolution, in situ data quickly at low cost (Hummel et al., 1996; Adamchuk et al., 2004), leading to greater overall accuracy in mapping soil variability. Lab-based analyses have two sources of error: i) analysis error due to subsampling and analytical determination, and ii) sampling error due to point-to-point variation in soils. With traditional soil testing, analysis error is expected to be relatively low, but sampling error can be substantial due to limited sampling density across space and time. Sensors can provide a spatiotemporal sampling intensity several orders of magnitude greater than traditional methods, thus reducing sampling error and potentially reducing overall error, even if analysis errors are higher (*e.g.*, Schirrmann et al., 2011).

Currently, sensors are widely used to monitor soil moisture content, salinity, and temperature, but sensors can also be used to estimate a range of inherent and dynamic soil attributes (Adamchuk et al., 2011; Wetterlind et al., 2015; Veum et al., 2016) related to soil health. Being able to reliably estimate soil health indicators in the field has clear benefits for sustainable agricultural management and environmental protection, but no single sensor has been identified or demonstrated to have the ability to provide accurate, meaningful data across the wide range of soil properties represented by soil health. Therefore, the current and widespread demand for soil health assessment represents an ideal opportunity for development and application of sensor fusion technologies.

Rapid, inexpensive, high resolution data is needed to facilitate field-scale soil health assessment across soils, regions, and climates, yet current measurements of soil health indicators generally involve field collection of soil samples followed by laboratory analyses that are not amenable to real-time monitoring of production fields. For on-farm assessments, methods that can efficiently map soil health across fields and landscapes and provide appropriate guidelines for interpretation are needed. Currently available indices for interpretation of laboratory soil health data, including the Soil Management and Assessment Framework (SMAF) and the Comprehensive Assessment of Soil Health (CASH), were developed with limited datasets. Both provided a good foundation for soil health assessment and new efforts are underway to revise and improve interpretive scoring curves for agricultural management decisions. This chapter will: (i) highlight novel tools and strategies for improved laboratory and in-field soil health data collection, (ii) identify challenges and critical gaps in current technology, and (iii) describe initiatives to improve science-based interpretation of soil health indicators for sustainable agriculture.

## Sensor Technology

Sensor technology has been employed in analytical chemistry, biology, and physics in laboratory settings since the early 20th century (Hardy 1938; Andrew 1984; Griffiths 2008). Many of these techniques still remain in the laboratory, but several have also been transferred to the field. Many in-field, sensor-based approaches are designed to capture data in a field setting without collecting soil samples or conducting laboratory analysis. By definition, proximal soil sensors are deployed in the field and are close to or in contact with the soil (Viscarra Rossel and McBratney 1998). In short, proximal soil sensing (PSS) brings the sensor to the soil in its native environment. These sensor techniques are generally based on the same fundamental chemical and physical principles as laboratory methods, with adaptations to handle field conditions. This in-field approach offers many advantages over traditional techniques and simultaneously faces unique challenges.

For several decades, PSS techniques have been used for traditional soil characterization (Viscarra Rossel et al., 2016), soil survey and mapping (Viscarra Rossel et al., 2010), and precision agriculture (Adamchuk et al., 2011). More recently, these techniques have been used to rapidly measure multiple soil health indicators and provide soil health index scores (Kinoshita et al., 2012; Veum et al., 2015b, 2017). The most common technique is to sense soil using diffuse reflectance spectroscopy with visible (400–700 nm), near-infrared [NIR; 750 to 2500 nm (13500 to 4000 $cm^{-1}$)] or mid-infrared [MIR; 2500 to 23500 nm (4000 to 450 cm-1)] ranges of the electromagnetic spectrum (Soriano Disla et al., 2014; Nocita et al., 2015; Johnson et al., 2019). This method is based on the interaction of light with soil constituents, producing spectral features that can subsequently be used to develop chemometric models for estimation of soil properties (Malley et al., 2004).

Statistical analysis of spectral datasets often requires multivariate techniques such as partial least squares regression (Viscarra Rossel and Behrens 2010) that can handle highly correlated, high-dimensional data where the number of prediction variables is greater than the number of observations (Haaland and Thomas 1988). More recently, Veum et al. (2018) applied Bayesian covariate assisted techniques to estimate SOC, total nitrogen (TN), and clay, silt, and sand content. In addition, a range of mathematical pretreatments or spectral transformations may be required, with no consensus on best practices in the literature (Stenberg et al., 2010). Likewise, no single calibration algorithm, including partial least squares regression, neural networks, or machine learning algorithms, have consistently provided best results (Igne et al., 2010; Viscarra Rossel and Behrens 2010; Pei et al., 2019). Additional sensors that have exhibited potential for in-field estimation of soil properties include apparent electrical conductivity (ECa; Corwin and Lesch 2005; Harvey and Morgan 2009), $\gamma$ radiation (Dierke and Werban 2013; Viscarra Rossel et al., 2017), and ground penetrating radar (Doolittle and Collins 1995). The application of these techniques to soil biological, physical, and chemical properties will be discussed in more detail below.

## Soil Biological Properties

Within the developing science of soil health, soil biology is playing a leading role in advancing our understanding of ecosystem function as well as agricultural productivity and sustainability (Lehman et al., 2015). Therefore, it is critical to gain a better understanding of soil biological function across space and time. Traditional laboratory methods for soil biological properties range from standard organic matter (OM) and soil organic carbon (SOC) measurements, to carbon and nitrogen fractions (e.g., Brookes et al., 1985; Doran 1987; Franzluebbers et al., 1996), to soil enzyme assays (e.g., Eivazi and Tabatabai 1988; Dick 1994). More recent advances in biological measurements include refinement of a rapid permanganate oxidation test (Weil et al., 2003; Culman et al., 2012), a combined enzyme panel (Acosta-Martinez et al., 2018; Acosta-Martinez et al., 2019), a rapid total protein extraction (Hurisso et al., 2018), and molecular analyses that determine microbial community composition via genomics (Manter et al., 2017) or high-throughput lipid analysis (Buyer and Sasser 2012).

Advancements for in-field sensing of soil biological indicators include a handheld DNA sequencer and portable DNA-extraction machine for virus detection (Shaffer 2019), a portable, real-time polymerase chain reaction (PCR) machine for on-site pathogen analysis (DeShields et al., 2018), and a variety of capacitance biosensors used in enzyme assays that provide label-free, real-time detection of antibodies, antigens, toxins, biomarkers, proteins, or nucleic acids (Bergdahl et al., 2019). Successful deployment and implementation of these techniques in the field is still an area of research and development.

Diffuse reflectance spectroscopy has been successfully used to estimate SOC or OM for some time in the laboratory as an alternative to wet chemistry methods (see reviews by Viscarra Rossel et al., 2006; Stenberg et al., 2010). In addition to SOC and OM, spectra in this range have been used to successfully estimate several other biological soil health indicators, including total nitrogen, β-glucosidase activity, active carbon, microbial biomass-carbon, microbial lipids, and soil respiration (Sudduth and Hummel 1993; Pietikäinen and Fritze 1995; Chang et al., 2001; Zornoza et al., 2008; Chaudhary et al., 2012; Kinoshita et al., 2012; Veum et al., 2014; Veum et al., 2015b), although estimation of those soil properties may be achieved by proxy through correlation with the spectral signature of SOC or OM. For example, good estimates of soil properties may be obtained by detecting changes in surrogate (*i.e.*, proxy or highly correlated) soil properties instead of directly detecting changes in the property of interest when using NIR spectroscopy methods (Vågen et al., 2006; Reeves 2010). Organic carbon bonds and inorganic mineral bonds are the primary sources of spectral features (see Viscarra Rossel et al., 2006; Cécillon et al., 2009), and when a model relies on an indirect, proxy measurement, performance may be inconsistent and will likely decline if the relationship between the target soil property and the proxy measurement is decoupled for any number of reasons.

## Soil Physical Properties

Physical soil properties, such as bulk density, macroaggregate stability, and water-filled pore space, reflect multiple soil functions including infiltration capacity, resistance to erosion, and microbial habitat (Angers 1992; Franzluebbers 1999; Logsdon and Karlen 2004). Laboratory measurement of these properties typically does not require expensive or sophisticated instrumentation, but the procedures are often time consuming and may require careful sample collection and handling procedures (e.g., Kemper and Koch 1966; Cambardella and Elliott 1992). Simple and rapid in-field aggregate tests have been proposed (Herrick et al., 2001; Seybold and Herrick 2001), but most in-field soil physical measurements, such as infiltration, pose difficulties due to high spatial variability and the extended time required for data collection (Haws et al., 2004; Angulo-Jaramillo et al., 2016).

In-field estimation of these important physical soil health indicators using reflectance spectra has been less consistent than for soil biological properties, although some studies have demonstrated success with soil texture and aggregate stability (Chang et al., 2001; Bogrekci and Lee 2005; Vågen et al., 2006) and bulk density (Cho et al., 2017a). Other in-field methods for soil physical properties include soil strength (Hemmat and Adamchuk 2008), $EC_a$ (Corwin and Lesch 2005), and ground penetrating radar (Doolittle and Collins 1995).

Soil strength, or soil penetration resistance (PR), is commonly measured as cone index by cone penetrometer and is defined as the force per unit base area required to push the penetrometer vertically through a specified increment of soil (ASAE, 2005; Chung et al., 2006). Included as a Tier 1 indicator for the Soil Health Institute assessment (soilhealthinstitute.org), PR is highly correlated with soil physical properties, such as bulk density, macroporosity, aggregate stability, moisture, degree of compaction, and soil texture (Chung et al., 2006; Nunes et al., 2019b), and has implications for important outcomes such as root growth (Nunes et al., 2019a). Horizontal, on-the-go soil strength sensors have also been developed to improve data collection efficiency (Chung et al., 2006; Hemmat and Adamchuk 2008).

Soil electrical conductivity is affected by numerous attributes, including soil texture, mineralogy, cation exchange capacity, and moisture (McNeill 1992; Sudduth et al., 2013; Doolittle and Brevik 2014). In non-saline soils, texture variations generally have the largest effect, so $EC_a$ has been used to map soil texture differences (Kitchen et al., 1996; Anderson-Cook et al., 2002). However, lack of a unique relationship between $EC_a$ values and specific soil properties, mainly due to the multiple properties that can affect $EC_a$ in any given situation, has hindered its use for quantitative assessment of soil properties in a practical setting. While good results are often obtained in research, it is difficult to elucidate protocols that will transfer those results to practice.

Ground penetrating radar has also been used since 1978 for soil survey purposes and to determine soil horizonation based on characteristics related to soil texture (Kung and Lu 1993; Doolittle and Collins 1995). In this way, GPR provides information on the subsurface location of variable soil physical properties, rather than quantitation of the property itself. Similarly, GPR was able to detect the depth to a compacted, higher bulk density layer but was not successful in assessing the relative density of the compacted zone (Raper et al., 1990).

## Chemical Properties

Chemical soil health indicators primarily reflect nutrient availability for crop growth and are typically represented by laboratory measurements of soil pH, electrical conductivity (paste), extractable P, and extractable K (Smith and Doran 1996; Staggenborg et al., 2007; Wienhold et al., 2009). Precision, in-field soil health assessments could provide high resolution information for improved management decisions, including fertilizer inputs (*e.g.*, right source, right rate, right time, right place), and would enhance the sustainability of production systems and reduce environmental impacts. However, in-field sensing of available P and K using reflectance spectra has met with limited success. A few researchers (e.g., Daniel et al., 2003; Bogrekci and Lee 2005) reported consistently good ($R^2 > 0.8$) results for both P and K, but other studies (e.g., Viscarra Rossel et al., 2006; Ge et al., 2007; Lee et al., 2009) reported consistently poor ($R^2 < 0.5$) results. This lack of consistency has been attributed to the fact that spectral estimation of P and K relies on covariation of nutrient concentrations with other, optically active soil constituents (Stenberg et al., 2010).

Most sensing of available P and K or pH has been by electrochemical means, with ion-selective membranes attached to electrodes (e.g., Kim, Hummel, and Birrell, 2006; Kim et al., 2007a) or to field-effect transistors (e.g., Birrell and Hummel, 2000). Sensing through direct soil contact has been less successful (Adamchuk et al., 2005) than when testing soil slurries (Sethuramasamyraja et al., 2008) or extracts (Kim et al., 2007b). Many of the studies of electrochemical sensing have been limited to the laboratory. A few systems for mobile, in-field sampling and quantification have been reported (Birrell and Hummel 2001; Sibley et al., 2009), yet none have been developed into commercial products. A major limitation is the difficulty implementing the steps required to acquire a sample and create a soil slurry or extract while traversing a field at a reasonable speed. A commercial system for on-the-go pH sensing was developed by Veris Technologies (Salina, KS) (Christy et al., 2004), and evaluated under field conditions by Schirrmann et al. (2011) who then provided operational guidance.

In contrast to an on-the-go system, another approach used in electrochemical sensing is a stationary or stop-and-go system. Smolka et al. (2017) evaluated this type of system for edge-of-field use based on its capability to measure $NO_3$, $NH_4$, K, and $PO_4$ on the basis of capillary electrophoresis. While $NO_3$ results were acceptable in field tests, they recommended further testing to confirm functionality for sensing other ions. Adamchuk et al. (2014) developed a stop-and-go system that moves to predetermined locations within a field, removes the surface soil layer, introduces multiple ISEs to the soil, and records sensor data along with geographic coordinates. In field tests, $NO_3$ and pH were strongly related to laboratory measurements, with $R^2$ of 0.87 and 0.84, respectively. Overall, development of in-field soil nutrient sensors remains an area of ongoing research.

## Auxiliary Data and Sensor Data Fusion

Soil health reflects an integration of physical, chemical, and biological functions, and it is not currently feasible to provide a complete soil health assessment with a single soil sensing approach. In particular, estimation models for important chemical and physical aspects of soil health have been less successful than the diffuse reflectance spectroscopy models for biological indicators. One technique to improve estimation of inconsistent target soil health indicators is to incorporate into the estimation models auxiliary variables that may be directly or indirectly related to the target soil property or may be less expensive to measure. This can be accomplished by combining simple laboratory measurements with reflectance spectra (Brown et al., 2006; Morgan et al., 2009; Kinoshita et al., 2012; Veum et al., 2015b; Veum et al., 2018). Although this may overcome some of the limitations of using reflectance spectra alone, this approach still requires collection and analysis of soil in the laboratory.

Another approach combines or 'fuses' data from complementary sensors and avoids laboratory soil analysis altogether (Kusumo et al., 2008; Adamchuk et al., 2011; Roudier et al., 2015; Cho et al., 2017b; Veum et al., 2017). A combination of reflectance spectra with data from electrochemical sensors may improve estimates of chemical and nutrient indicators. For example, La et al. (2016) found that P and K estimates were improved when combining ISE and NIR spectral data ($R^2 \geq 0.95$) compared with ISE data alone ($R^2 \geq 0.87$). The increased accuracy was attributed to the ability of the spectral data to provide an estimate of soil texture. In another example, an infield core-scanning system that included γ-ray attenuation and digital imaging along with NIR spectroscopy was used to estimate multiple soil profile properties (Viscarra Rossel et al., 2017). In addition, the combination of reflectance spectra with $EC_a$ and penetration (cone index) sensor data has demonstrated improved estimates of multiple soil properties (Veum et al., 2017; Pei et al., 2019). These studies illustrate the potential for rapid quantification of soil health by fusing auxiliary laboratory data or data obtained from complementary sensors. However, critical gaps remain in the development and application of sensors for nutrient availability, soil structure, and advanced soil biological indicators.

## Challenges to In situ Data Collection

Several challenges to infield sensor applications have been identified (Ben-Dor et al., 2009). It has been routinely shown that reflectance spectra are sensitive to soil moisture and other environmental conditions (*i.e.*, temperature and soil structure), thus decreasing the estimation accuracy of spectral data and hampering the potential use of spectra collected in the field (Sudduth and Hummel 1993; Morgan et al., 2009; Reeves 2010; Minasny et al., 2011). Specifically, the O–H bands from soil moisture mask important wavelengths produced by organic carbon bonds and other important functional groups (Mouazen et al., 2007). To mitigate this problem, Sudduth and Hummel (1993) suggested including a wide range of soil moisture contents in calibration models. Even in the laboratory, soil handling and pretreatment, such as soil sample drying, grinding, or sieving protocols, can impact spectral features and affect the accuracy of model estimates. Several studies have compared soils in a field-moist condition or soil samples rewetted to prescribed moisture levels with air or oven-dried soil (Sudduth and Hummel 1993; Fystro 2002; Stevens et al., 2006; Waiser et al., 2007; Morgan et al., 2009; Nocita et al., 2013). Across studies, the effect of soil moisture is inconsistent, but overall, dry soil samples generally provides more robust results.

Several techniques have been applied to in situ NIR spectra to account for moisture and other environmental factors to improve model performance. These include external parameter orthogonalization (EPO), direct standardization, and global moisture modeling (Wijewardane et al., 2016). The EPO algorithm projects the soil spectra orthogonal to the space of unwanted variation (Roger et al., 2003). Studies have successfully used EPO to estimate SOC (Minasny et al., 2011; Ge et al., 2014; Veum et al., 2018) and clay content (Ge et al., 2014; Ackerson et al., 2017; Veum et al., 2018). Alternatively, the direct standardization approach derives a transfer matrix that has successfully been used to predict OM using a portable spectrometer (Ji et al., 2015). For global moisture modeling, a secondary variable with a relationship to the primary variable is intentionally manipulated to improve the calibration model (Kawano et al., 1995). Similar to spiking, this approach has been applied to datasets across geographical regions and to spectral libraries for estimation of soil carbon and clay content (Brown et al., 2006; Wetterlind and Stenberg 2010), with the goal of developing models that are robust to variable environmental conditions.

## Soil Profile Information

Soil health research has focused on surface soils (around 0 to 15 cm) that are critical for plant growth and development and represent the zone of primary agricultural management. Surface soils are also the interface with the atmosphere, and alteration of the upper soil horizons through management can impact many biological, physical, and chemical properties and processes (Hursh and Hoover, 1942). In addition, there are practical limitations in soil sampling equipment and budget constraints that result in an emphasis on surface soil horizons in most soil health assessments. However, vertical stratification and anisotropy in the soil profile affects infiltration, storage, and movement of water (Bouma et al., 1977), nutrients (Jobbágy and Jackson 2001), carbon storage (Harrison et al., 2011), and microbial activity (Fontaine et al., 2007). For example, the subsurface argillic horizon characteristic of claypan soils impacts water partitioning, runoff, lateral movement of water through the soil, as well as microbial community structure and function (Hsiao et al., 2018). Vertical stratification of soil health indicators is also frequently observed under no-till (NT) systems. Adoption of NT is one of the most important strategies for conservation agriculture and improvement of soil health (Veum et al., 2014; Nunes et al., 2018). However, long-term NT managed without other conservation practices (*e.g.*, diversified cropping systems and permanent soil cover by mulching and/or cover crops) can also result in strong stratification of chemical, physical, and biological soil health indicators, with a higher concentration of nutrients and SOC in the uppermost topsoil layer (~0 to 8-cm), and lower nutrient availability with increased soil compaction below that depth (Powlson and Jenkinson 1981; Deubel et al., 2011; Houx et al., 2011; Nunes et al., 2019a). These characteristics may restrict plant root growth, decrease the absorption and translocation of water and nutrients to plants, and ultimately limit crop yield. Furthermore, soil compaction also reduces water infiltrability (infiltration capacity), which can induce soil erosion by water, transport pesticides and nutrients to water bodies, and contribute to environmental pollution. Therefore, in-field, profile estimation of soil properties could provide vertical spatial data that would be useful in understanding soil function and site-specific constraints (*e.g.*, subsurface soil compaction) that are essential for informed management and enhanced crop production.

Soil profile characteristics can be measured on deep core samples returned to the laboratory (Hummel et al., 2001; Lee et al., 2009) or post-extraction on intact cores in the field (Kusumo et al., 2008; Roudier et al., 2015). Acquisition of in situ profile soil information offers many potential benefits over laboratory analysis of profile samples in terms of efficiency, timeliness, and expense and has been investigated by several groups. Figure 8.1 illustrates the type of data a commercial, in situ profile sensor (Veris P4000) is capable of simultaneously collecting VNIR spectra, $EC_a$, and PR to a depth of 1 m (Christy et al., 2011).

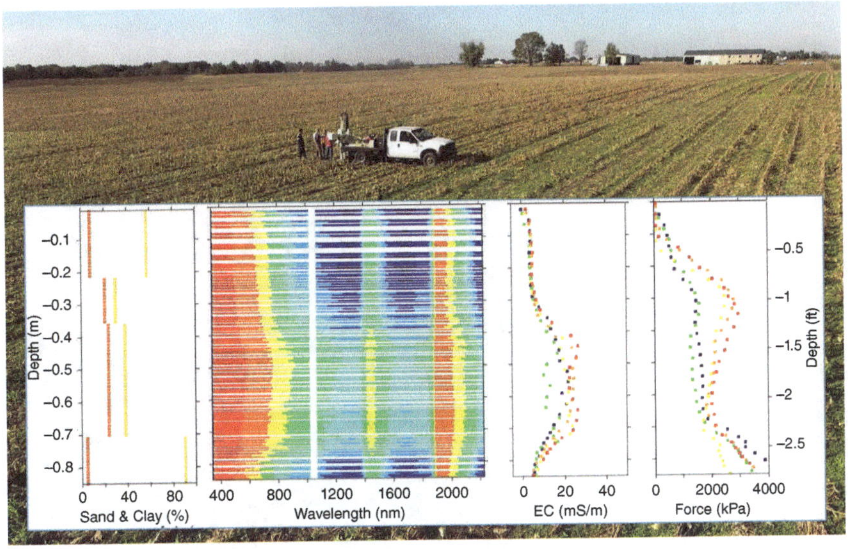


**Figure 8.1** Example of in situ profile estimation of soil properties using VNIR reflectance, apparent electrical conductivity, and penetration resistance data, with laboratory sand and clay content (image from Karlen et al., 2019).

Data from the P4000 instrument has been used to estimate multiple profile soil properties. Kweon et al. (2008) successfully used this probe in situ for SOC estimation, Wetterlind et al. (2015) successfully estimated texture and OM, and Cho et al. (2017b) obtained moderate to low cross-validation $R^2$ for bulk density, SOC, soil moisture, clay, silt, and sand. Pei et al. (2019) fused the three P4000 data streams and improved estimations of several physical and chemical soil properties compared with VNIR spectra alone; however only calcium improved by more than 5%. Hodge and Sudduth (2012) compared the SOC estimation accuracy of this spectrometer on dry samples scanned in the laboratory with in situ field data collected with the same instrument, finding that the in situ data was less accurate. These results were similar to the findings of Veum et al. (2018), and the reduction in accuracy is likely due to the aforementioned environmental effects associated with in situ data collection.

As an alternative to direct measurement depthwise through the soil profile, it is possible to infer depth distributions of soil $EC_a$ from mobile sensor data collection. If two or more $EC_a$ measurements with different depth response functions are available, then mathematical inversion of the theoretical models of the $EC_a$ instrument response can reconstruct the EC in different layers of the soil profile (Monteiro Santos et al., 2010). Those layer EC values can then be calibrated to layer soil properties as demonstrated by Sudduth et al. (2017) for soil texture (Fig. 8.2).

**Figure 8.2** Example of mapping clay content at multiple depths through fusion of mobile (lateral) apparent electrical conductivity ($EC_a$) data and profile (vertical) $EC_a$ data (image from Karlen et al., 2019).

## Soil Health Indices

Advancing soil health assessment also depends on improvements in the interpretation of soil health indicators (Karlen et al., 2019). In the past decades, several soil health assessment and monitoring tools were developed (Bünemann et al., 2018), including the SMAF, developed by Andrews et al. (2004), and the CASH, developed by Cornell University (Moebius-Clune et al., 2016). The SMAF employs multiple dynamic chemical, physical, and biological soil health indicator data to assess management effects on soil functions using a three-step process which includes indicator selection, interpretation, and integration into a final soil health index. The interpretation process is based on nonlinear scoring curves. These curves were developed to be sensitive to multiple site-specific factors, including soil texture, soil taxonomy (soil suborder), clay mineralogy, slope, climate (precipitation and temperature regimes), sampling timing (season), and analytical measurement technique, among other factors. Currently, SMAF has scoring curves (interpretation algorithms) for four biological soil properties (SOC, microbial biomass-C, potentially mineralizable-N, and β-glucosidase activity); four physical soil properties (bulk density, macroaggregate stability, available water capacity, and water-filled pore space); and five chemical soil properties (pH, electrical conductivity, sodium adsorption ratio, extractable P, and extractable K) (Andrews et al., 2004; Wienhold et al., 2009; Stott et al., 2011). The CASH, developed by the Cornell Soil Health Testing Laboratory, offers soil health testing packages to landowners and provides interpretive management guidance together with the laboratory results. The CASH interpretative scoring curves, based primarily on the principle of cumulative normal distribution (Moebius-Clune et al., 2016), were initially developed with a large dataset from the Northeastern United States, and were recently expanded to include soils from the Midwestern United States (Fine et al., 2017

Since its release in 2004, the SMAF has been used internationally to compare tillage, rotation, amendment, and grazing practices (e.g., Cambardella et al., 2004; Karlen et al., 2008; Ozgoz et al., 2013; Stott et al., 2013; Karlen and Johnson 2014; Veum et al., 2014; Gelaw et al., 2015; Veum et al., 2015a; Cherubin et al., 2016). However, the raw dataset used to develop the SMAF scoring curves was limited in scope, with an emphasis on data from three different U.S. locations (*i.e.*, Iowa, California, and Georgia).

Ideally, interpretation curves are based on a broad range of data, where threshold (*e.g.*, maximum, minimum, optimum) values can be well defined to provide appropriate scoring functions. Karlen et al. (2019) documented the evolution of the soil health concept over the past several decades and suggested that soil health assessment tools, such as SMAF, should be improved. This involves development of protocols for national (and eventually global) soil health monitoring, as well as identification and calibration of new indicators of soil biological, chemical, and physical soil health. Thus, the USDA Natural Resources Conservation Service (NRCS) and Agricultural Research Service (ARS) initiated a nationwide SMAF meta-analysis to address that need by critically evaluating and improving the current SMAF algorithms and incorporating scoring functions for an expanded suite of soil health indicators.

To this end, a broad national database was developed including soil health indicator data from 456 published studies conducted across 38 states from several U.S. regions (Fig. 8.3a). Most of the soil health indicators included in the database are from topsoil observations (Fig. 8.3b). Following SOC, chemical soil health indicators were the most sampled followed by physical and biological properties (Fig. 8.4). As expected, more than 60% of the papers presented SOC as the primary soil health indicator. In addition, the metadata represented several soil types (*e.g.*, 10 soil orders, 31 soil suborders, 11 soil texture), varying climate conditions, and several land use and management practices, including addition of amendments, different tillage intensities, and cropping systems.

This nationwide assessment confirmed that the multi-functional biological, physical, and chemical soil health indicators used in SMAF and CASH are

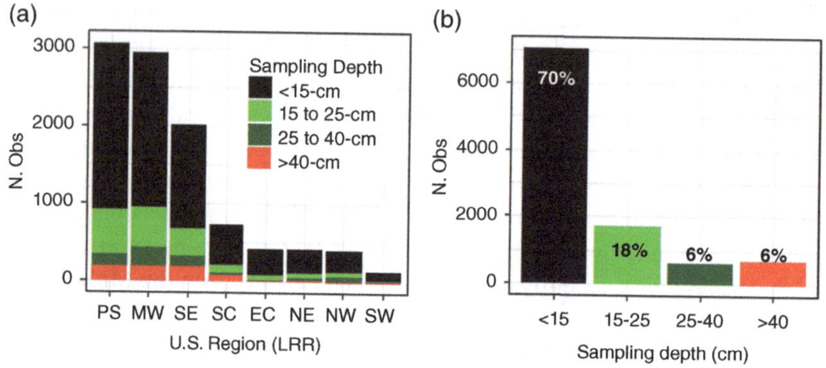


**Figure 8.3** Number of soil health observations by U.S. region (a) and sampling depth (b). where: PS = Plains States (ND, KS, NE, SD, MT, WY, CO); MW = Midwest (IA, IL, IN, MN, MO, WI, MW, OH); SE = Southeast (FL, NC, SC, GA, AL, MS, DC, NJ, DE, MD, VA); SC = South Central (TX, NM, OK, AR, LA); EC = East Central (PA, WV, KY, TN); NE = Northeast (NY, CT, MA, ME, NH, RI, VT, MI); NW = Northwest (WA, OR, ID); and SW = Southwest (CA, AZ, UT, NV).

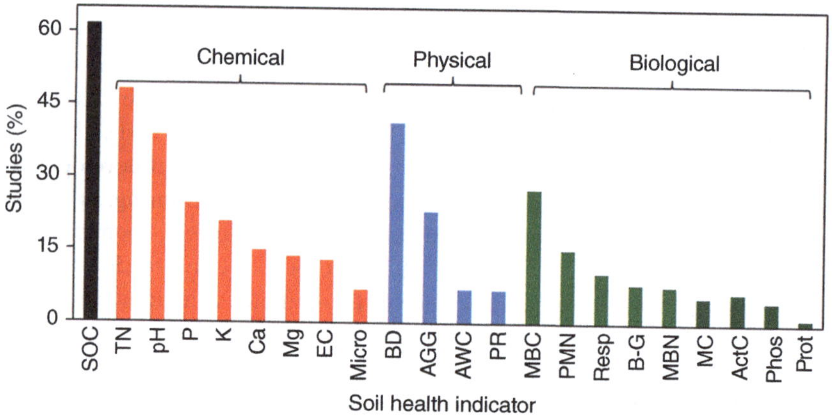


**Figure 8.4** Percentage of studies that assessed soil organic carbon (SOC), chemical indicator data [total nitrogen (TN), pH, exchangeable P, K, Ca, Mg, electrical conductivity (EC), and micronutrients (Micro)], soil physical properties [bulk density (BD), aggregate stability (AGG), available water capacity (AWC), and penetration resistance (PR)], and soil biological indicator data [microbial biomass carbon (MBC), potential mineralizable nitrogen (PMN), soil respiration (Resp), β-glucosidase (B-G), microbial biomass nitrogen (MBN), mineralizable carbon (MC), active carbon (ActC), phosphatase activity (Phos), and soil protein (Prot)].

sensitive to changes induced by agronomic management practices and land-use decisions, independent of the U.S. region, which is in accordance with past studies (Zornoza et al., 2008; Veum et al., 2015a; Nunes et al., 2018; Van Es and Karlen 2019). In addition, the meta-analysis results confirmed that the SMAF scoring curves are sensitive to and can detect soil health changes induced by land use and management practices; however, this analysis suggested that a few SMAF algorithms need to be closely evaluated for potential revision. As noted in other studies, the current SMAF algorithms may be underestimating SOC and β-glucosidase scores (Stott et al., 2013; Mbuthia et al., 2015; Zobeck et al., 2015) and potentially overestimating aggregate stability scores (Stott et al., 2011; Veum et al., 2015a). Therefore, the Soil Health Assessment Protocol and Evaluation (SHAPE) was recently developed to account for climate and edaphic factors and include estimates of uncertainty (Nunes et al., 2021). Access to the SHAPE Shiny app or GitHub repository is available upon request from Kristen Veum.

Achieving the goal of rapid, in-field assessment of soil health will also require the ability to provide comprehensive maps of soil function, or 'SHAPE maps' gleaned from in-field, sensor estimates of soil health indicators. Ideally, the soil health indicator data will be transformed via on-the-go software into interpretive maps for producers. Thus far, few studies have attempted the simultaneous estimation of a broad range of biological, physical, and chemical soil properties using PSS techniques for the purpose of a comprehensive soil health assessment. Vågen et al. (2006) estimated a three-category soil fertility index based on 10 indicators, Cohen et al. (2006) classified three levels of ecological soil degradation using 17 indicators, and Kinoshita et al. (2012) estimated a three-category soil quality index based on the CASH. Others have related NIR spectra to more general soil quality indices. For example, Cécillon et al. (2009) successfully modeled three indicators related to OM, nutrient supply, and biological activity in a forest ecosystem, and Paz-Kagan et al. (2014) developed a spectral soil quality index that discriminated between broad land use categories and related well to a laboratory-based soil quality index. In contrast, Veum et al. (2015b) estimated SMAF soil health indicators and continuous (as opposed to discrete categorical) scores by combining spectra with auxiliary laboratory data. They found that NIR worked well for estimation of the biological components of the SMAF and performed poorly for estimation of chemical or physical SMAF scores. A subsequent study by Veum et al. (2017) demonstrated improved estimates of the physical component of the SMAF soil health assessment when using a sensor data fusion approach over using NIR spectra alone. Nonetheless, these studies have often drawn samples from a wide geographic area or from non-agricultural areas such as forests, and few studies have investigated the ability of soil sensors to estimate multiple soil health indicators and scores at the field scale. More research is needed to improve in-field estimation of soil health indicators, particularly soil chemical properties, and a refined interpretation index is needed before this can be achieved.

### Conclusions and Future Work

Measuring multiple soil health indicators at high temporal or spatial frequency is time consuming and costly using conventional laboratory methods. Furthermore, within the developing science of soil health, soil biological and physical function are playing a leading role in advancing our understanding of ecosystem function as well as agricultural productivity and sustainability. Therefore, it is critical to gain a better understanding of soil health across space and time, and alternative approaches are needed to achieve high resolution, field scale information for producers. Development of infield sensors will facilitate collection of high-resolution data that will expand our knowledge of soil variability to improve crop production and protect soil and water resources. To accomplish the goal of in situ measurements that rapidly quantify soil health, integration of advanced sensor platforms and data analysis techniques will almost certainly be required. Despite many successes in this area to date, there remain important research and development needs, including new sensor technology to more directly and accurately measure soil structure and strength, soil nutrients, and advanced biological characteristics. Sensor optimization to improve data quality and reliability is needed, and methods to address a range of environmental conditions will be necessary to move additional sensor techniques from the laboratory to the field.

The new SHAPE interpretations will provide improved guidance for landowners and producers. Increased accessibility to soil health information and management recommendations will allow producers and landowners to make informed, science-based decisions. The integration of infield sensing data with SHAPE scoring curves in user-friendly software applications could provide multi-layer soil health maps as a cost-effective decision tool for landowners. As soil health is adopted at the national and global scale, interpretative algorithms will also need to be expanded to include regionally relevant soil and climate characteristics and indicators. Ultimately, data-driven soil health management information that is ecologically relevant is needed to optimize management decisions at the field scale to provide economic and environmental benefits.

## References

Ackerson, J.P., Morgan, C.L.S., and Ge, Y. 2017. Penetrometer-mounted VisNIR spectroscopy: Application of EPO-PLS to in situ VisNIR spectra. *Geoderma* 286, 131–138. doi:10.1016/j.geoderma.2016.10.018

Acosta-Martinez, V., Cano, A., and Johnson, J. (2018). Simultaneous determination of multiple soil enzyme activities for soil health-biogeochemical indices. *Appl. Soil Ecol.* 126, 121–128. doi:10.1016/j.apsoil.2017.11.024

Acosta-Martinez, V., Pérez-Guzmán, L., and Johnson, J. (2019). Simultaneous determination of β-glucosidase, β-glucosaminidase, acid phosphomonoesterase, and arylsulfatase activities in a soil sample for a biogeochemical cycling index. *Appl. Soil Ecol.* 142, 72–80. doi:10.1016/j.apsoil.2019.05.001

Adamchuk, V.I., Dhawale, N., and Rene-Laforest, F. (2014). Development of an on-the-spot analyser for measuring soil chemical properties. In: *Proceedings of the 12th International Conference on Precision Agriculture*. Monticello, Illinois: International Society of Precision Agriculture.

Adamchuk, V.I., Hummel, J.W., Morgan, M.T., and Upadhyaya, S.K. (2004). On-the-go soil sensors for precision agriculture. *Comput. Electron. Agric.* 44, 71–91. doi:10.1016/j.compag.2004.03.002

Adamchuk, V.I., Lund, E.D., Sethuramasamyraja, B., Morgan, M.T., Dobermann, A., and Marx, D.B. (2005). Direct measurement of soil chemical properties on-the-go using ion-selective electrodes. *Comput. Electron. Agric.* 48(3), 272–294. doi:10.1016/j.compag.2005.05.001

Adamchuk, V.I., R.A. Viscarra Rossel, K.A. Sudduth, and P. Schulze Lammers. (2011). Sensor fusion for precision agriculture. In: C. Thomas (Ed.), *Sensor Fusion- Foundation and Applications* (p. 27–40). Rijeka, Croatia: InTech.

Anderson-Cook, C.M., Alley, M.M., Roygard, J.K.F., and Khosla, R. (2002). Differentiating soil types using electromagnetic conductivity and crop yield maps. *Soil Sci. Soc. Am. J.* 66, 1562–1570. doi:10.2136/sssaj2002.1562

Andrew, E.R. (1984). A historical review of NMR and its clinical applications. *Br. Med. Bull.* 40(2), 115–119. doi:10.1093/oxfordjournals.bmb.a071956

Andrews, S.S., Karlen, D.L., and Cambardella, C.A. 2004. The Soil Management Assessment Framework: A quantitative soil quality evaluation method. *Soil Sci. Soc. Am. J.* 68, 1945–1962. doi:10.2136/sssaj2004.1945

Angers, D.A. 1992. Changes in soil aggregation and organic carbon under corn and alfalfa. *Soil Sci. Soc. Am. J.* 56:1244–1249. doi:10.2136/sssaj199 2.03615995005600040039x

Angulo-Jaramillo, R., Bagarello, V., Iovino, M., and Lassabatere, L. (2016). *Infiltration Measurements for Soil Hydraulic Characterization*. Amsterdam, The Netherlands: Springer International Publishing. doi:10.1007/978-3-319-31788-5

ASAE. (2005). S313.3: Soil cone penetrometer In: American Society of Agricultural Engineers, editor, *ASAE Standards*. American Society of Agricultural Engineers, St. Joseph, MI.

Bakhshandeh, E., Francaviglia, R., and Renzi, G. (2019). A cost and time-effective method to evaluate soil microbial respiration for soil quality assessment. *Appl. Soil Ecol.* 140, 121–125. doi:10.1016/j.apsoil.2019.04.023

Ben-Dor, E., Chabrillat, S., Demattê, J.A., Taylor, G.R., Hill, J., Whiting, M., and Sommer, S. (2009). Using imaging spectroscopy to study soil properties. *Remote Sens. Environ.* 113, S38–S55.

Bergdahl, G.E., Hedström, M., and Mattiasson, B. (2019). Capacitive sensor to monitor enzyme activity by following degradation of macromolecules in real time. *Appl. Biochem. Biotechnol.* 189(2), 374–383. doi:10.1007/s12010-019-03006-0

Birrell, S.J., and Hummel, J.W. (2000). Membrane selection and ISFET configuration evaluation for soil nitrate sensing. *Trans. ASAE* 43(2), 197–206. doi:10.13031/2013.2694

Birrell, S.J., and Hummel, J.W. (2001). Real-time multi-ISFET/FIA soil analysis system with automatic sample extraction. *Comput. Electron. Agric.* 32(1), 45–67. doi:10.1016/S0168-1699(01)00159-4

Bogrekci, I., and Lee, W.S. (2005). Improving phosphorus sensing by eliminating soil particle size effect in spectral measurement. *Trans. ASAE* 48(5), 1971–1978. doi:10.13031/2013.19989

Bouma, J., Jongerius, A., Boersma, O., Jager, A., and Schoonderbeek, D. (1977). The function of different types of macropores during saturated flow through four swelling soil horizons. *Soil Sci. Soc. Am. J.* 41(5), 945–950. doi:10.2136/sssaj1977.03615995004100050028x

Brookes, P.C., Landman, A., Pruden, G., and Jenkinson, D.S. (1985). Chloroform fumigation and the release of soil nitrogen: A rapid direct extraction method to measure microbial biomass nitrogen in soil. *Soil Biol. Biochem.* 17(6), 837–842. doi:10.1016/0038-0717(85)90144-0

Brown, D.J., Shepherd, K.D., Walsh, M.G., Mays, M.D., and Reinsch, T.G. (2006). Global soil characterization with VNIR diffuse reflectance spectroscopy. *Geoderma* 132(3-4), 273–290. doi:10.1016/j.geoderma.2005.04.025

Bünemann, E.K., Bongiornio, G., Bai, Z., Creamer, R.E., de Deyn, G., de Goede, R., Fleskens, L., Geissen, V., Kuyper, T.W., Mader, P., Pulleman, M., Sukkel, W., Van Groeningen, J.W., and Brussaard, L.T. (2018). Soil qualtiy- a critical review. *Soil Biol. Biochem.* 120, 105–125. doi:10.1016/j.soilbio.2018.01.030

Buyer, J.S., and Sasser, M. (2012). High throughput phospholipid fatty acid analysis of soils. *Appl. Soil Ecol.* 61, 127–130. doi:10.1016/j.apsoil.2012.06.005

Cambardella, C., Moorman, T.B., Andrews, S.S., and Karlen, D.L. (2004). Watershed-scale assessment of soil quality in the loess hills of southwest Iowa. *Soil Tillage Res.* 78, 237–247. doi:10.1016/j.still.2004.02.015

Cambardella, C.A., and Elliott, E.T. (1992). Particulate soil organic-matter changes across a grassland cultivation sequence. *Soil Sci. Soc. Am. J.* 56(3), 777–783. doi:10.2136/sssaj1992.03615995005600030017x

Cécillon, L., Barthès, B.G., Gomez, C., Ertlen, D., Genot, V., Hedde, M., Stevens, A., and Brun, J.J. (2009). Assessment and monitoring of soil quality using near-infrared reflectance spectroscopy (NIRS). *Eur. J. Soil Sci.* 60(5), 770–784. doi:10.1111/j.1365-2389.2009.01178.x

Chang, C.W., Laird, D.A., Mausbach, M.J., and Hurburgh, C.R. (2001). Near-infrared reflectance spectroscopy-principal components regression analysis of soil properties. *Soil Sci. Soc. Am. J.* 65, 480–490. doi:10.2136/sssaj2001.652480x

Chaudhary, V.P., Sudduth, K.A., Kitchen, N.R., and Kremer, R.J. 2012. Reflectance spectroscopy detects management and landscape differences in soil carbon and nitrogen. *Soil Sci. Soc. Am. J.* 76(2), 597–606. doi:10.2136/sssaj2011.0112

Cherubin, M.R., Karlen, D.L., Franco, A.L.C., Cerri, C.E.P., Tormena, C.A., and Cerri, C.C. (2016). A Soil Management Assessment Framework (SMAF) evaluation of Brazilian sugarcane expansion on soil quality. *Soil Sci. Soc. Am. J.* 80(1), 215–226. doi:10.2136/sssaj2015.09.0328

Cho, Y., Sheridan, A.H., Sudduth, K.A., and Veum, K.S. (2017a). Comparison of field and laboratory VNIR spectroscopy for profile soil property estimation. *Trans. ASABE* 60(5), 1503–1510. doi:10.13031/trans.12299

Cho, Y., Sudduth, K.A., and Drummond, S.T. (2017b). Profile soil property estimation using a VIS-NIR-EC-force probe. *Trans. ASABE* 60(3), 683–692. doi:10.13031/trans.12049

Christy, C., Drummond, P., Kweon, G., Maxton, C., Drelling, K., Jensen, K., and Lund, E. (2011). *Multiple sensor system and method for mapping soil in three dimensions.* U.S. Patent No. US 20110106451 A1.

Christy, C.D., Collings, K., Drummond, P., and Lund, E. 2004. *A mobile sensor platform for measurement of soil pH and buffering*. ASAE Paper No. 041042. St. Joseph, MI: ASABE. doi:10.13031/2013.16138

Chung, S.O., Sudduth, K.A., and Hummel, J.W. (2006). Design and validation of an on-the-go soil strength profile sensor. *Trans. ASABE* 49(1), 5–14. doi:10.13031/2013.20229

Cohen, M.J., Dabral, S., Graham, W.D., Prenger, J.P., and Debusk, W.F. (2006). Evaluating ecological condition using soil biogeochemical parameters and near infrared reflectance spectra. *Environ. Monit. Assess.* 116(1-3), 427–457. doi:10.1007/s10661-006-7664-8

Corwin, D.L., and Lesch, S.M. (2005). Apparent soil electrical conductivity measurements in agriculture. *Comput. Electron. Agric.* 46(1–3), 11–43. doi:10.1016/j.compag.2004.10.005

Culman, S.W., Snapp, S.S., Freeman, M.A., Schipanski, M.E., Beniston, J., Lal, R., Drinkwater, L.E., Franzluebbers, A.J., Glover, J.D., Grandy, A.S., Lee, J., Six, J., Maul, J.E., Mirsky, S.B., Spargo, J.T., and Wander, M.M. (2012). Permanganate oxidizable carbon reflects a processed soil fraction that is sensitive to management. *Soil Sci. Soc. Am. J.* 76(2), 494–504. doi:10.2136/sssaj2011.0286

Daniel, K.W., Tripathi, N.K., and Honda, K. (2003). Artificial neural network analysis of laboratory and in situ spectra for the estimation of macronutrients in soils of Lop Buri (Thailand). *Aust. J. Soil Res.* 41(1), 47–59. doi:10.1071/SR02027

DeForest, J.L. (2009). The influence of time, storage temperature, and substrate age on potential soil enzyme activity in acidic forest soils using MUB-linked substrates and L-DOPA. *Soil Biol. Biochem.* 41(6), 1180–1186. doi:10.1016/j.soilbio.2009.02.029

DeShields, J.B., Bomberger, R.A., Woodhall, J.W., Wheeler, D.L., Moroz, N., Johnson, D.A., and Tanaka, K. (2018). On-site molecular detection of soil-Bborne phytopathogens using a portable real-time PCR system. *J. Vis. Exp.* 2018(132), 56891.

Deubel, A., Hofmann, B., and Orzessek, D. (2011). Long-term effects of tillage on stratification and plant availability of phosphate and potassium in a loess chernozem. *Soil Tillage Res.* 117, 85–92. doi:10.1016/j.still.2011.09.001

Dick, R.P. (1994). Soil enzyme activity as an indicator of soil quality. In J.W. Doran, D.C. Coleman, B.A. Stewart, and D.F. Bezdicek (Eds.), *Defining soil quality for a sustainable environment* (p. 107–124). SSSA Spec. Publ. No. 35. Madison, WI: SSSA, ASA. doi:10.2136/sssaspecpub35.c7

Dierke, C., and Werban, U. (2013). Relationships between gamma-ray data and soil properties at an agricultural test site. *Geoderma* 199, 90–98. doi:10.1016/j.geoderma.2012.10.017

Doolittle, J.A., and Brevik, E.C. (2014). The use of electromagnetic induction techniques in soils studies. *Geoderma* 223-225, 33–45. doi:10.1016/j.geoderma.2014.01.027

Doolittle, J.A., and Collins, M.E. (1995). Use of soil information to determine application of ground penetrating radar. *J. Appl. Geophys.* 33(1–3), 101–108. doi:10.1016/0926-9851(95)90033-0

Doran, J.W. (1987). Microbial biomass and mineralizable nitrogen distributions in no-tillage and plowed soils. *Biol. Fertil. Soils* 5(1), 68–75. doi:10.1007/BF00264349

Doran, J.W., and Parkin, T.B. (1996). Quantitative indicators of soil quality: A minimum data set. In J.W. Doran, and A.J. Jones (Eds.), *Methods for Assessing Soil Quality* (p. 25–37). Madison, WI: Soil Science Society of America.

Doran, J.W., and Zeiss, M.R. (2000). Soil health and sustainability: Managing the biotic component of soil quality. *Appl. Soil Ecol.* 15, 3–11. doi:10.1016/S0929-1393(00)00067-6

Eivazi, F., and Tabatabai, M.A. (1988). Glucosidases and galactosidases in soils. *Soil Biol. Biochem.* 20(5), 601–606. doi:10.1016/0038-0717(88)90141-1

Fine, A. K., van Es, H. M., & Schindelbeck, R. R. (2017). Statistics, scoring functions, and regional analysis of a comprehensive soil health database. *Soil Science Society of America Journal*, 81(3), 589–601. doi:10.2136/sssaj2016.09.0286

Fontaine, S., Barot, S., Barré, P., Bdioui, N., Mary, B., and Rumpel, C. (2007). Stability of organic carbon in deep soil layers controlled by fresh carbon supply. *Nature* 450:277. doi:10.1038/nature06275

Franzluebbers, A., and Veum, K.S. (2020). Comparison of two alkali trap methods for measuring the flush of $CO_2$. *Agron. J.* doi:10.1002/agj2.20141

Franzluebbers, A.J. (1999). Microbial activity in response to water-filled pore space of variably eroded southern Piedmont soils. *Appl. Soil Ecol.* 11(1), 91–101. doi:10.1016/S0929-1393(98)00128-0

Franzluebbers, A.J., Haney, R.L., Hons, F.M., and Zuberer, D.A. (1996). Determination of microbial biomass and nitrogen mineralization following rewetting of dried soil. *Soil Sci. Soc. Am. J.* 60(4), 1133–1139. doi:10.2136/sssaj1996.03615995006000040025x

Fystro, G. (2002). The prediction of C and N content and their potential mineralization in heterogeneous soil samples using VIS-NIR spectroscopy and comparative methods. *Plant Soil* 246(2), 139–149. doi:10.1023/A:1020612319014

Ge, Y., Morgan, C.L.S., and Ackerson, J.P. (2014). VisNIR spectra of dried ground soils predict properties of soils scanned moist and intact. *Geoderma* 221–222, 61–69. doi:10.1016/j.geoderma.2014.01.011

Ge, Y., Thomasson, J.A., Morgan, C.L., and Searcy, S.W. (2007). VNIR diffuse reflectance spectroscopy for agricultural soil property determination based on regression-kriging. *Trans. ASABE* 50(3), 1081–1092. doi:10.13031/2013.23122

Gelaw, A., Singh, B., and Lal, R. (2015). Soil quality indices for evaluating smallholder agricultural land uses in Northern Ethiopia. *Sustainability* 7(3), 2322. doi:10.3390/su7032322

Giacometti, C., Cavani, L., Baldoni, G., Ciavatta, C., Marzadori, C., and Kandeler, E. (2014). Microplate-scale fluorometric soil enzyme assays as tools to assess soil quality in a long-term agricultural field experiment. *Appl. Soil Ecol.* 75, 80–85. doi:10.1016/j.apsoil.2013.10.009

Griffiths, J. (2008). A brief history of mass spectrometry. *Anal. Chem.* 80(15), 5678–5683. doi:10.1021/ac8013065

Haaland, D.M., and Thomas, E.V. (1988). Partial least-squares methods for spectral analyses. 1. Relation of other quantitative calibration methods and the extraction of qualitative information. *Anal. Chem.* 60(11), 1193–1202. doi:10.1021/ac00162a020

Haney, R., Hons, F.M., Sanderson, M., and Franzluebbers, A. (2001). A rapid procedure for estimating nitrogen mineralization in manured soil. *Biol. Fertil. Soils* 33, 100–104. doi:10.1007/s003740000294

Hardy, A.C. (1938). History of the design of the recording spectrophotometer. *J. Opt. Soc. Am.* 28, 360–364. doi:10.1364/JOSA.28.000360

Harrison, R., Footen, P., and Strahm, B. (2011). Deep soil horizons: Contribution and importance to soil carbon pools and in assessing whole-ecosystem response to management and global change. *For. Sci.* 57, 67–76.

Harvey, O.R., and Morgan, C.L.S. (2009). Predicting regional-scale soil variability using a single calibrated apparent soil electrical conductivity model. *Soil Sci. Soc. Am. J.* 73(1), 164–169.

Haws, N.W., Liu, B., Boast, C.W., Rao, P.S.C., Kladivko, E.J., and Franzmeier, D.P. (2004). Spatial variablity on measurement scale of infiltration rate on an agricultural landscape. *Soil Sci. Soc. Am. J.* 68, 1818–1826. doi:10.2136/sssaj2004.1818

Hemmat, A., and Adamchuk, V.I. (2008). Sensor systems for measuring soil compaction: Review and analysis. *Comput. Electron. Agric.* 63(2), 89–103. doi:10.1016/j.compag.2008.03.001

Herrick, J.E., Whitford, W.G., de Soyza, A.G., Van Zee, J.W., Havstad, K.M., Seybol, C.A., and Walton, M. (2001). Field soil aggregate stbility kit for soil quality and rangeland health evaluation. *Catena* 44, 27–35. doi:10.1016/S0341-8162(00)00173-9

Hodge, A.M., and Sudduth, K.A. (2012). *Comparison of two spectrometers for profile soil carbon sensing.* ASABE Paper No. 121338240. St. Joseph, MI: ASABE.

Houx, J.H., Wiebold, W.J., and Fritschi, F.B. (2011). Long-term tillage and crop rotation determines the mineral nutrient distributions of some elements in a Vertic Epiaqualf. *Soil Tillage Res.* 112, 27–35. doi:10.1016/j.still.2010.11.003

Hsiao, C.-J., Sassenrath, G.F., Zeglin, L.H., Hettiarachchi, G.M., and Rice, C.W. (2018). Vertical changes of soil microbial properties in claypan soils. *Soil Biol. Biochem.* 121, 154–164. doi:10.1016/j.soilbio.2018.03.012

Hummel, J.W., Gaultney, L.D., and Sudduth, K.A. (1996). Soil property sensing for site-specific crop management. *Comput. Electron. Agric.* 14, 121–136. doi:10.1016/0168-1699(95)00043-7

Hummel, J.W., Sudduth, K.A., and Hollinger, S.E. (2001). Soil moisture and organic matter prediction of surface and subsurface soils using an NIR soil sensor. *Comput. Electron. Agric.* 32, 149–165. doi:10.1016/S0168-1699(01)00163-6

Hurisso, T.T., Moebius-Clune, D.J., Culman, S.W., Moebius-Clune, B.N., Thies, J.E., and van Es, H.M. 2018. Soil protein as a rapid soil health indicator of potentially available organic nitrogen. *Agric. Environ. Lett.* 3(1), 1–5.

Hursh, C.R., and Hoover, M.D. (1942). Soil profile characteristics pertinent to hydrologic studies in the Southern Appalachains. *Soil Sci. Soc. Am. J.* 6, 414–422. doi:10.2136/sssaj1942.036159950006000C0077x

Igne, B., Reeves, III, J.B., McCarty, G., Hively, W.D., Lund, E., and Hurburgh, Jr, C.R. (2010). Evaluation of spectral pretreatments, partial least squares, least squares support vector machines and locally weighted regression for quantitative spectroscopic analysis of soils. *Near Infrared Spectroscopy* 18, 167–176.

Janik, L.J., Merry, R.H., and Skjemstad, J.O. (1998). Can mid infrared diffuse reflectance analysis replace soil extractions? *Aust. J. Exp. Agric.* 38, 681–696. doi:10.1071/EA97144

Ji, W., Viscarra Rossel, R.A., and Shi, Z. (2015). Accounting for the effects of water and the environment on proximally sensed vis–NIR soil spectra and their calibrations. *Eur. J. Soil Sci.* 66(3), 555–565. doi:10.1111/ejss.12239

Jobbágy, E., and Jackson, R. (2001). The distribution of soil nutrients with depth: Global patterns and the imprint of plants. *Biogeochemistry* 53, 51–77. doi:10.1023/A:1010760720215

Johnson, J.-M., Vandamme, E., Senthilkumar, K., Sila, A., Shepherd, K.D., and Saito, K. (2019). Near-infrared, mid-infrared or combined diffuse reflectance spectroscopy for assessing soil fertility in rice fields in sub-Saharan Africa. *Geoderma* 354, 113840. doi:10.1016/j.geoderma.2019.06.043

Jung, W.K., Kitchen, N.R., Sudduth, K.A., and Anderson, S.H. (2006). Spatial characteristics of claypan soil properties in an agricultural field. *Soil Sci. Soc. Am. J.* 70, 1387–1397. doi:10.2136/sssaj2005.0273

Karlen, D., and Johnson, J.F. (2014). Crop residue considerations for sustainable bioenergy feedstock supplies. *BioEnergy Res.* 7(2), 465–467. doi:10.1007/s12155-014-9407-y

Karlen, D.L., Mausbach, M.J., Doran, J.W., Cline, R.G., Harris, R.F., and Schuman, G.E. (1997). Soil quality: A concept, definition, and framework for evaluation. *Soil Sci. Soc. Am. J.* 61, 4–10. doi:10.2136/sssaj1997.03615995006100010001x

Karlen, D.L., Tomer, M.D., Neppel, J., and Cambardella, C. 2008. A preliminary watershed scale soil quality assessment in north central Iowa, USA. *Soil Tillage Res.* 99, 291–299. doi:10.1016/j.still.2008.03.002

Karlen, D.L., Veum, K.S., Sudduth, K.A., Obrycki, J.F., and Nunes, M.R. (2019). Soil health assessment: Past accomplishments, current activities, and future opportunities. *Soil Tillage Res.* 195, 104365.

Kawano, S., Abe, H., and Iwamoto, M. (1995). Development of a calibration equation with temperature compensation for determining the Brix value in intact peaches. *J. Near Infrared Spectrosc.* 3(4), 211–218. doi:10.1255/jnirs.71

Kemper, W.D., and Koch, E.J. (1966). *Aggregate stability of soils from Western United States and Canada; measurement procedure, correlations with soil constituents.* United States. Dep. of Agriculture Technical bulletin no. 1355 No. 52. Washington, D.C.: Agricultural Research Service, U.S. Dep. of Agriculture.

Kim, H.J., Hummel, J.W., and Birrell, S.J. (2006). Evaluation of nitrate and potassium ion-selective membranes for soil macronutrient sensing. *Trans. ASABE* 49(3), 597–606. doi:10.13031/2013.20476

Kim, H.J., Hummel, J.W., Sudduth, K.A., and Birrell, S.J. (2007a). Evaluation of phosphate ion-selective membranes and cobalt-based electrodes for soil nutrient sensing. *Trans. ASABE* 50(2), 215–225.

Kim, H.J., Hummel, J.W., Sudduth, K.A., and Motavalli, P.P. (2007b). Simultaneous analysis of soil macronutrients using ion-selective electrodes. *Soil Sci. Soc. Am. J.* 71(6),1867–1877. doi:10.2136/sssaj2007.0002

Kinoshita, R., Moebius-Clune, B.N., van Es, H.M., Hively, W.D., and Bilgilis, A.V. (2012). Strategies for soil quality assessment using visible and near-infrared reflectance spectroscopy in a Western Kenya chronosequence. *Soil Sci. Soc. Am. J.* 76(5), 1776–1788. doi:10.2136/sssaj2011.0307

Kitchen, N.R., Sudduth, K.A., and Drummond, S.T. (1996). Mapping of sand deposition from 1993 Midwest floods with electromagnetic induction measurements. *J. Soil Water Conserv.* 51(4), 336–340.

Kung, K.-J.S., and Lu, Z.-B. (1993). Using ground-penetrating radar to detect layers of discontinuous dielectric constant. *Soil Sci. Soc. Am. J.* 57(2), 335–340. doi:10.2136/sssaj1993.03615995005700020008x

Kusumo, B.H., Hedley, C.B., Hedley, M.J., Hueni, A., Tuohy, M.P., and Arnold, G.C. (2008). The use of diffuse reflectance spectroscopy for in situ carbon and nitrogen analysis of pastoral soils. *Aust. J. Soil Res.* 46(7), 623–635. doi:10.1071/SR08118

Kweon, G., Lund, E., Maxton, C., Drummond, P., and Jensen, K. 2008. *In situ measurement of soil properties using a probe based VIS-NIR spectrophotometer.* ASABE Paper No. 084399. St. Joseph, Michigan: American Society of Agricultural and Biological Engineers.

La, W.J., Sudduth, K.A., Kim, H.J., and Chung, S.O. (2016). Fusion of spectral and electrochemical sensor data for estimating soil macronutrients. *Trans. ASABE* 59(4), 787–794. doi:10.13031/trans.59.11562

Lee, K.S., Lee, D.H., Sudduth, K.A., Chung, S.O., Kitchen, N.R., and Drummond, S.T. (2009). Wavelength identification and diffuse reflectance estimation for surface and profile soil properties. *Trans. ASABE* 52, 683–695. doi:10.13031/2013.27385

Lehman, R., Cambardella, C., Stott, D., Acosta-Martinez, V., Manter, D., Buyer, J., Maul, J., Smith, J., Collins, H., Halvorson, J., Kremer, R., Lundgren, J., Ducey, T., Jin, V., and Karlen, D. (2015). Understanding and enhancing soil biological health: The solution for reversing soil degradation. *Sustainability* 7(1), 988.

Logsdon, S.D., and Karlen, D.L. (2004). Bulk density as a soil quality indicator during conversion to no-tillage. *Soil Tillage Res.* 78(2), 143–149. doi:10.1016/j.still.2004.02.003

Malley, D.F., Martin, P.D., and Ben-Dor, E. (2004). Application in analysis of soils. In C.A. Roberts, J. Workman Jr, and J.B. Reeves III (Eds.), *Near-Infrared Spectroscopy in Agriculture* (p.729–784). Madison, WI: American Society of Agronomy, Crop Science Society of America, Soil Science Society of America. doi:10.2134/agronmonogr44.c26

Manter, D.K., Delgado, J.A., Blackburn, H.D., Harmel, D., Pérez de León, A.A., and Honeycutt, C.W. 2017. Opinion: Why we need a national living soil repository. *Proc. Natl. Acad. Sci. USA* 114(52), 13587–13590. doi:10.1073/pnas.1720262115

Mbuthia, L.W., Acosta-Martínez, V., DeBruyn, J., Schaeffer, S., Tyler, D., Odoi, E., Mpheshea, M., Walker, F., and Eash, N. 2015. Long term tillage, cover crop, and fertilization effects on microbial community structure, activity: Implications for soil quality. *Soil Biol. Biochem.* 89, 24–34. doi:10.1016/j.soilbio.2015.06.016

McNeill, J.D. 1992. Rapid, accurate mapping of soil salinity by electromagnetic ground conductivity meters. In: G.C. Topp, W.D. Reynolds, and R.E. Green, editors, *Advances in measurement of soil physical properties: Bringing theory into practice.* SSSA, Madison, WI. doi:10.2136/sssaspecpub30.c11

Minasny, B., McBratney, A.B., Bellon-Maurel, V., Roger, J.-M., Gobrecht, A., Ferrand, L., and Joalland, S. (2011). Removing the effect of soil moisture from NIR diffuse reflectance spectra for the prediction of soil organic carbon. *Geoderma* 167–168, 118–124. doi:10.1016/j.geoderma.2011.09.008

Moebius-Clune, B.N., D.J. Moebius-Clune, B.K. Gugino, O.J. Idowu, R.R. Schindelbeck, A.J. Ristow, H.M. van Es, J.E. Thies, H.A. Shayler, M.B. McBride, D.W. Wolfe, and G.S. Abawi. 2016. *Comprehensive Assessment of Soil Health: The Cornell Framework Manual.* 3.1 ed. Ithaca, NY: Cornell University.

Monteiro Santos, F.A., Triantafilis, J., Bruzgulis, K.E., and Roe, J.A.E. (2010). Inversion of multiconfiguration electromagnetic (DUALEM-421) profiling data using a one-dimensional laterally constrained algorithm. *Vadose Zone J.* 9, 117–125. doi:10.2136/vzj2009.0088

Morgan, C.L.S., Waiser, T.H., Brown, D.J., and Hallmark, C.T. (2009). Simulated in situ characterization of soil organic and inorganic carbon with visible near-infrared diffuse reflectance spectroscopy. *Geoderma* 151(3–4), 249–256. doi:10.1016/j.geoderma.2009.04.010

Mouazen, A. M., Maleki, M. R., De Baerdemaeker, J., & Ramon, H. (2007). On-line measurement of some selected soil properties using a VIS–NIR sensor. *Soil and Tillage Research*, 93(1), 13–27. doi:http://dx.doi.org/10.1016/j.still.2006.03.009

Nocita, M., Stevens, A., Noon, C., and van Wesemael, B. (2013). Prediction of soil organic carbon for different levels of soil moisture using Vis-NIR spectroscopy. *Geoderma* 199(0), 37–42. doi:10.1016/j.geoderma.2012.07.020

Nocita, M., Stevens, A., van Wesemael, B., Aitkenhead, M., Bachmann, M., Barthès, B., Ben Dor, E., Brown, D.J., Clairotte, M., Csorba, A., Dardenne, P., Demattê, J.A.M., Genot, V., Guerrero, C., Knadel, M., Montanarella, L., Noon, C., Ramirez-Lopez, L., Robertson, J., Sakai, H., Soriano-Disla, J.M., Shepherd, K.D., Stenberg, B., Towett, E.K., Vargas, R., and Wetterlind, J. 2015. Chapter four- Soil spectroscopy: An alternative to wet chemistry for soil monitoring. In D.L. Sparks (Ed.), *Advances in Agronomy* (p. 139–159). Waltham, MA: Academic Press.

Nortcliff, S. (2002). Standardization of soil quality attributes. *Agric. Ecosyst. Environ.* 88, 161–168. doi:10.1016/S0167-8809(01)00253-5

Nunes, M., van Es, H., Schindelbeck, R., James Ristow, A., and Ryan, M. (2018). No-till and cropping system diversification improve soil health and crop yield. *Geoderma* 328, 30–43. doi:10.1016/j.geoderma.2018.04.031

Nunes, M.R., Karlen, D.L., Denardin, J.E., and Cambardella, C.A. (2019a). Corn root and soil health indicator response to no-till production practices. *Agric. Ecosyst. Environ.* 285,106607.

Nunes, M.R., Pauletto, E.A., Denardin, J.E., Suzukid, L.E.A.S., and van Es, H.M. (2019b). Dynamic changes in compressive properties and crop response after chisel tillage in a highly weathered soil. *Soil Tillage Res.* 186, 183–190. doi:10.1016/j.still.2018.10.017

Nunes, M. R., Veum, K. S., Parker, P. A., Holan, S. H., Wills, S., Seybold, C. A., Van Es, H. M., Amsili, J., Karlen, D., & Moorman, T. B. (2021). The Soil Health Assessment Protocol and Evaluation (SHAPE) applied to soil organic C. *Soil Science Society of America Journal*, https://doi.org/10.1002/saj2.20244

Ozgoz, E., Gunal, H., Acir, N., Gokmen, F., Birol, M., and Budak, M. (2013). Soil quality and spatial variability in assessment of land use effects in a typic haplustoll. *Land Degrad. Dev.* 24(3), 277–286. doi:10.1002/ldr.1126

Pachepsky, Y., and M. Van Genuchten. 2011. Pedotransfer functions. In J. Gliński, J. Horabik, J. Lipiec (Eds.), *Encyclopedia of agrophysics* (p. 556–561). Amsterdam, The Netherlands.

Paz-Kagan, T., Shachak, M., Zaady, E., and Karnieli, A. (2014). A spectral soil quality index (SSQI) for characterizing soil function in areas of chaged land use. *Geoderma* 230-231, 171–184. doi:10.1016/j.geoderma.2014.04.003

Pei, X., Sudduth, K.A., Veum, K.S., and Li, M. (2019). Improving in-situ estimation of soil profile properties using a multi-sensor probe. *Sensors (Basel Switzerland)* 19(5), 1011. doi:10.3390/s19051011

Pietikäinen, J., and Fritze, H. (1995). Clear-cutting and prescribed burning in coniferous forest: Comparison of effects on soil fungal and total microbial biomass, respiration activity and nitrification. *Soil Biol. Biochem.* 27(1), 101–109. doi:10.1016/0038-0717(94)00125-K

Piotrowska-Długosz, A., Breza-Boruta, B., and Długosz, J. (2019). Spatial and temporal variability of the soil microbiological properties in two soils with a different pedogenesis cropped to winter rape (*Brassica napus* L.). *Geoderma* 340, 313–324. doi:10.1016/j.geoderma.2019.01.020

Powlson, D.S., and Jenkinson, D.S. (1981). A comparison of the organic matter, biomass, adenosine triphosphate and mineralizable nitrogen contents of ploughed and direct-drilled soils. *J. Agric. Sci.* 97(3), 713–721. doi:10.1017/S0021859600037084

Raper, R.L., Asmussen, L.E., and Powell, J.B. (1990). Sensing hardpan depth with ground-penetrating radar. *Trans. ASAE* 33(1), 41–46. doi:10.13031/2013.31291

Reeves, J.B. (2010). Near- versus mid-infrared diffuse reflectance spectroscopy for soil analysis emphasizing carbon and laboratory versus on-site analysis: Where are we and what needs to be done? *Geoderma* 158(1–2,:3–14. doi:10.1016/j.geoderma.2009.04.005

Roger, J.M., Chauchard, F., and Bellon Maurel, V. (2003). EPO-PLS external parameter orthogonalisation of PLS application to temperature-independent measurement of sugar content of intact fruits. *Chemom. Intell. Lab. Syst.* 66(2), 191–204. doi:10.1016/S0169-7439(03)00051-0

Roudier, P., Hedley, C.B., and Ross, C.W. 2015. Prediction of volumetric soil organic carbon from field-moist intact soil cores. *Eur. J. Soil Sci.* 66, 651–660. doi:10.1111/ejss.12259

Schirrmann, M., Gebbers, R., Kramer, E., and Seidel, J. (2011). Soil pH mapping with an on-the-go sensor. *Sensors (Basel Switzerland)* 11(1), 573–598. doi:10.3390/s110100573

Schomberg, H.H., Wietholter, S., Griffin, T.S., Reeves, D.W., Cabrera, M.L., Fisher, D.S., Endale, D.M., Novak, J.M., Balkcom, K.S., Raper, R.L., Kitchen, N.R., Locke, M.A., Potter, K.N., Schwartz, R.C., Truman, C.C., and Tyler, D.D. (2009). Assessing indices for predicting potential nitrogen mineralization in soils under different management systems. *Soil Sci. Soc. Am. J.* 73(5), 1575–1586. doi:10.2136/sssaj2008.0303

Sethuramasamyraja, B., Adamchuk, V.I., Dobermann, A., Marx, D.B., Jones, D.D., and Meyer, G.E. (2008). Agitated soil measurement method for integrated on-the-go mapping of soil pH, potassium, and nitrate contents. *Comput. Electron. Agric.* 60(2), 212–225. doi:10.1016/j.compag.2007.08.003

Seybold, C.A., and Herrick, J.E. (2001). Aggregate stability kit for soil quality assessments. *Catena* 44, 37–45. doi:10.1016/S0341-8162(00)00175-2

Shaffer, L. (2019). Inner workings: Portable DNA sequencer helps farmers stymie devastating viruses. *Proc. Natl. Acad. Sci. USA* 116(9), 3351–3353. doi:10.1073/pnas.1901806116

Sibley, K.J., Astatkie, T., Brewster, G., Struik, P.C., Adsett, J.F., and Pruski, K. (2009). Field-scale validation of an automated soil nitrate extraction and measurement system. *Precis. Agric.* 10(2), 162–174. doi:10.1007/s11119-008-9081-1

Smith, J.L., and Doran, J.W. (1996). Measurement and use of pH and electrical conductivity for soil quality analysis. In J.W. Doran, and A.J. Jones (Eds.), *Methods for assessing soil quality* (p. 169–185). Madison, WI: Soil Science Society of America, Inc.

Smolka, M., Puchberger-Enengl, D., Bipoun, M., Klasa, A., Kiczkajlo, M., Śmiechowski, W., Sowinski, P., Krutzler, C., Keplinger, F., and Vellekoop, M.J. (2017). A mobile lab-on-a-chip device for on-site soil nutrient analysis. *Precis. Agric.* 18, 152–168. doi:10.1007/s11119-016-9452-y

Soriano Disla, J., Janik, L., Viscarra Rossel, R., Macdonald, L., and McLaughlin, M. (2014). The performance of visible, near-, and mid-infrared reflectance spectroscopy for prediction of soil physical, chemical, and biological properties. *Appl. Spectrosc. Rev.* 49, 139–186. doi:10.1080/05704928.2013.811081

Staggenborg, S.A., Carignano, M., and Haag, L. (2007). Predicting soil pH and buffer pH in situ with a real-time sensor. *Agron. J.* 99(3), 854–861. doi:10.2134/agronj2006.0254

Stenberg, B., Viscarra Rossel, R.A., Mouazen, A.M., and Wetterlind, J. (2010). Chapter five-Visible and near infrared spectroscopy in soil science. In D.L. Sparks (Ed.) *Advances in Agronomy* (p. 163–215). Waltham, MA: Academic Press.

Stevens, A., van Wesemael, B., Vandenschrick, G., Touré, S., and Tychon, B. (2006). Detection of carbon stock change in agricultural soils using spectroscopic techniques. *Soil Sci. Soc. Am. J.* 70(3), 844–850. doi:10.2136/sssaj2005.0025

Stine, M.A., and Weil, R.R. (2002). The relationship between soil quality and crop prodictivity across three tillage systems in south central Honduras. *Am. J. Altern. Agric.* 17(1), 2–8.

Stott, D.E. 2019. *Recommended soil health indicators and associated laboratory procedures.* Soil Health Technical Note No. 450-03. Washington, D.C.: U.S. Dep. Agriculture-NRCS.

Stott, D.E., Cambardella, C.A., Tomer, M.D., Karlen, D.L., and Wolf, R. (2011). A soil quality assessment within the Iowa River South Fork watershed. *Soil Sci. Soc. Am. J.* 75(6), 2271–2282. doi:10.2136/sssaj2010.0440

Stott, D.E., Karlen, D.L., Cambardella, C.A., and Harmel, R.D. (2013). A soil quality and metabolic activity assessment after fifty-seven years of agricultural management. *Soil Sci. Soc. Am. J.* 77(3), 903–913. doi:10.2136/sssaj2012.0355

Sudduth, K.A., and Hummel, J.W. (1993). Soil organic matter, CEC, and moisture sensing with a prototype NIR spectrometer. *Trans. ASAE* 36, 1571–1582. doi:10.13031/2013.28498

Sudduth, K.A., Kitchen, N.R. and Drummond, S.T. (2017). Inversion of soil electrical conductivity data to estimate layered soil properties. In: *Proceedings of the 11th European Conf. on Precision Agriculture, Advances in Animal Biosciences* (p. 433–438), Edinburgh, Scotland, 16-20 July. Cambridge, U.K.: Cambridge Univ. Press. doi:10.1017/S2040470017001303

Sudduth, K.A., Myers, D.B., Kitchen, N.R., and Drummond, S.T. (2013). Modeling soil electrical conductivity–depth relationships with data from proximal and penetrating ECa sensors. *Geoderma* 199, 12–21. doi:10.1016/j.geoderma.2012.10.006

Tóth, B., Weynants, M., Nemes, A., Makó, A., Bilas, G., and Tóth, G. (2015). New generation of hydraulic pedotransfer functions for Europe. *Eur. J. Soil Sci.* 66(1), 226–238.

Vågen, T.-G., Shepherd, K.D., and Walsh, M.G. (2006). Sensing landscape level change in soil fertility following deforestation and conversion in the highlands of Madagascar using Vis-NIR spectroscopy. *Geoderma* 133(3–4), 281–294. doi:10.1016/j.geoderma.2005.07.014

van Es, H.M., and Karlen, D. (2019). Reanalysis validates soil health indicator sensitivity and correlation with long-term crop yields. *Soil Sci. Soc. Am. J.* doi:10.2136/sssaj2018.09.0338

Veum, K.S., Goyne, K.W., Kremer, R.J., Miles, R.J., and Sudduth, K.A. (2014). Biological indicators of soil quality and soil organic matter characteristics in an agricultural management continuum. *Biogeochemistry* 117, 81–99. doi:10.1007/s10533-013-9868-7

Veum, K.S., Kremer, R.J., Sudduth, K.A., Kitchen, N.R., Lerch, R.N., Baffaut, C., Stott, D.E., Karlen, D.L., and Sadler, E.J. (2015a). Conservation effects on soil quality indicators in the Missouri Salt River Basin. *J. Soil Water Conserv.* 70(4), 232–246. doi:10.2489/jswc.70.4.232

Veum, K.S., Lorenz, T.L., and Kremer, R.J. (2019). Phospholipid fatty acid profiles of soils under variable handling and storage conditions. *Agron. J.* 111(3), 1090–1096. doi:10.2134/agronj2018.09.0628

Veum, K.S., Parker, P.A., Sudduth, K.A., and Holan, S.H. (2018). Predicting profile soil properties with reflectance spectra via Bayesian covariate assisted external parameter orthogonalization. *Sensors (Basel Switzerland)* 18(11), 3869. doi:10.3390/s18113869

Veum, K.S., Sudduth, K.A., and N.R. Kitchen. (2016). Sensor based soil health assessment. *Proceedings of the 13th International Society of Precision Agriculture, July 31-August 4, 2016. St. Louis, MO.* ASA, CSSA, SSSA, Madison, WI.

Veum, K.S., Sudduth, K.A., Kremer, R.J., and Kitchen, N.R. (2015b). Estimating a soil quality index with VNIR reflectance spectroscopy. *Soil Sci. Soc. Am. J.* 79(2), 637–649. doi:10.2136/sssaj2014.09.0390

Veum, K.S., Sudduth, K.A., Kremer, R.J., and Kitchen, N.R. (2017). Sensor data fusion for soil health assessment. *Geoderma* 305, 53–61. doi:10.1016/j.geoderma.2017.05.031

Viscarra Rossel, R.A., and Behrens, V. (2010). Using data mining to model and interpret soil diffuse reflectance spectra. *Geoderma* 158, 46–54. doi:10.1016/j.geoderma.2009.12.025

Viscarra Rossel, R.A., Behrens, T., Ben-Dor, E., Brown, D.J., Demattê, J.A.M., Shepherd, K.D., Shi, Z., Stenberg, B., Stevens, A., Adamchuk, V., Aïchi, H., Barthès, B.G., Bartholomeus, H.M., Bayer, A.D., Bernoux, M., Böttcher, K., Brodský, L., Du, C.W., Chappell, A., Fouad, Y., Genot, V., Gomez, C., Grunwald, S., Gubler, A., Guerrero, C., Hedley, C.B., Knadel, M., Morrás, H.J.M., Nocita, M., Ramirez-Lopez, L., Roudier, P., Campos, E.M.R., Sanborn, P., Sellitto, V.M., Sudduth, K.A., Rawlins, B.G., Walter, C., Winowiecki, L.A., Hong, S.Y., and Ji, W. (2016). A global spectral library to characterize the world's soil. *Earth Sci. Rev.* 155, 198–230. doi:10.1016/j.earscirev.2016.01.012

Viscarra Rossel, R.A., Lobsey, C.R., Sharman, C., Flick, P., and McLachlan, G. (2017). Novel proximal sensing for monitoring soil organic C stocks and condition. *Environ. Sci. Technol.* 51(10), 5630–5641. doi:10.1021/acs.est.7b00889

Viscarra Rossel, R.A., and McBratney, A.B. (1998). Laboratory evaluation of a proximal sensing technique for simultaneous measurement of soil clay and water content. *Geoderma* 85(1), 19–39. doi:10.1016/S0016-7061(98)00023-8

Viscarra Rossel, R.A., McKenzie, N.J., and Grundy, M.J. (2010). Using proximal soil sensors for digital soil mapping. In J.L. Boettinger, D.W. Howell, A.C. Moore, A.E. Hartemink, and S. Kienast-Brown (Eds.), *Digital soil mapping: Bridging research, environmental application, and operation* (p. 79–92). Dordrecht: Springer Netherlands.

Viscarra Rossel, R.A., Walvoort, D.J.J., McBratney, A.B., Janik, L.J., and Skjemstad, J.O. (2006). Visible, near infrared, mid infrared or combined diffuse reflectance spectroscopy for simultaneous assessment of various soil properties. *Geoderma* 131, 59–75. doi:10.1016/j.geoderma.2005.03.007

Waiser, T.H., Morgan, C.L.S., Brown, D.J., and Hallmark, C.T. (2007). In situ characterization of soil clay content with visible near-infrared diffuse reflectance spectroscopy. *Soil Sci. Soc. Am. J.* 71(2), 389–396. doi:10.2136/sssaj2006.0211

Weil, R.R., Islam, K.R., Stine, M.A., Gruver, J.B., and Samson-Liebig, S.E. (2003). Estimating active carbon for soil quality assessment: A simplified method for laboratory and field use. *Am. J. Altern. Agric.* 18(1):3–17. doi:10.1079/AJAA200228

Wetterlind, J., Piikki, K., Stenberg, B., and Söderström, M. (2015). Exploring the predictability of soil texture and organic matter content with a commercial integrated soil profiling tool. *Eur. J. Soil Sci.* 66(4), 631–638. doi:10.1111/ejss.12228

Wetterlind, J., and Stenberg, B. (2010). Near-infrared spectroscopy for within-field soil characterization: Small local calibrations compared with national libraries spiked with local samples. *Eur. J. Soil Sci.* 61(6), 823–843. doi:10.1111/j.1365-2389.2010.01283.x

Wienhold, B.J., Karlen, D.L., Andrews, S.S., and Stott, D.E. (2009). Protocol for indicator scoring in the soil management assessment framework (SMAF). *Renew. Agric. Food Syst.* 24(4), 260–266. doi:10.1017/S1742170509990093

Wijewardane, N.K., Ge, Y., and Morgan, C.L.S. (2016). Prediction of soil organic and inorganic carbon at different moisture contents with dry ground VNIR: A comparative study of different approaches. *Eur. J. Soil Sci.* 67(5), 605–615. doi:10.1111/ejss.12362

Zobeck, T.M., Steiner, J.L., Stott, D.E., Duke, S.E., Starks, P.J., Moriasi, D.N., and Karlen, D.L. (2015). Soil quality index comparisons using Fort Cobb, Oklahoma, watershed-scale land management data. *Soil Sci. Soc. Am. J.* 79, 224–238. doi:10.2136/sssaj2014.06.0257

Zornoza, R., Guerrero, C., Mataix-Solera, J., Scow, K.M., Arcenegui, V., and Mataix-Beneyto, J. (2008). Near infrared spectroscopy for determination of various physical, chemical and biochemical properties in Mediterranean soils. *Soil Biol. Biochem.* 40(7), 1923–1930. doi:10.1016/j.soilbio.2008.04.003

# Epilogue

*Douglas L. Karlen*

Soil is a ubiquitous, fragile resource that most people never think about unless there is dirt on their cell phones, mud on their trucks, or muck in their favorite rivers and lakes. Soil may also garner some attention when grass and shrubs fail to grow in local green space because of that "hard clay" or when they hear the world is running out of sand (Torres et al. 2018). But in reality, since the time of Plato (~5000 BCE), public recognition and appreciation of soil has been sparce. Such generalizations are not true for everyone, and as a career Soil Scientist it is comforting to know there have been U.S. presidents, scientists, and literary writers who have pleaded with humanity to care for soil (Table E.1).

I also find it interesting that several global soil proverbs (Yang et al., 2018) including many from our Native American brothers and sisters have embraced and stressed the intimate linkage between land, soil, water, air, and humanity for centuries (Table E.2). Without question, it has been said many times by many inhabitants on the American continent and around the entire world that we must appreciate and take care of our fragile soil resources as if our very lives depend upon it – because it's true!

A multitude of techniques have been used to increase general awareness of our fragile, living soil resources. This includes the successful soil exhibit entitled "Dig It" developed by the Smithsonian's National Museum of Natural History with support from the Soil Science Society of America (SSSA) and its lead sponsor, the Nutrients for Life Foundation, as well as Bayer CropScience, LI-COR Biosciences, Syngenta, and the USDA. The exhibit was not only on display for two years (2008 to 2010) in the Smithsonian National Museum of Natural History in Washington D.C., but then traveled around the U.S. for seven years before being placed on permanent display at the at the Saint Louis Science Center in St. Louis, MO.

*Soil Health Series: Volume 1 Approaches to Soil Health Analysis*, First Edition.
Edited by Douglas L. Karlen, Diane E. Stott, and Maysoon M. Mikha.

**Table E.1** Selected quotes advocating for recognition and better management of soil resources.

| Quote | Source |
|---|---|
| A nation that destroys its soil, destroys itself | Franklin Delano Roosevelt |
| A thin layer of earth, a few inches of rain, and a blanket of air, make human life possible on our planet | John F. Kennedy |
| Civilization itself rests upon the soil | Thomas Jefferson |
| There can be no life without soil and no soil without life, they have evolved together | Charles E. Kellogg |
| A new day in agriculture will come if, and when, we get both our hands and our minds a bit deeper into our soils | William A. Albrecht |
| Take care of the land and the land will take care of you | Hugh Hammond Bennett |
| The real wealth of the Nation lies in the resources of the earth, — soil, water, forests, minerals, and wildlife | Rachel Carson |
| Of celestial body movement, far more do we know, then about soil underfoot | Da Vinci |
| [Soil is] the thin layer covering the planet that stands between us and starvation | W. E. Larson |
| What we do to the land, we do to ourselves | Wendell Berry |
| We do not inherit the earth from our ancestors, we borrow it from our children | Chief Seattle |

**Table E.2** Selected Native American proverbs reflecting upon land and soil resources.

| Proverb | Source |
|---|---|
| The earth is the mother of all people,<br>All people should have equal rights upon it. | Attributed to Chief Joseph |
| Take only what you need,<br>Leave the land as you found it. | Attributed to the Arapaho |
| Mother Earth gave us an abundance of blessings,<br>To gather along life's path. | Unattributed Native American Proverb |
| Touching the Earth,<br>Equates to having harmony with nature. | Unattributed Native American Proverb |
| The ground on which we stand,<br>Is sacred ground. | Chief Plenty Coups, Crow Tribe |

Another activity to build public awareness of soil resources was establishment of official "State Soils." In 20 states, these have been legislatively established and share the same level of distinction as official State Flowers and Birds. In Wisconsin, grassroot efforts resulting in the designation of the Antigo silt loam as the official soil in 1983 were led by Dr. Francis D. Hole[1] (Wikipedia, 2020). As an expression of his love for soils, Dr. Hole wrote lyrics, played soft tunes on his fiddle, and sang songs about soil with students and others throughout his University of Wisconsin – Madison career (https://soils.wisc.edu/people/history/fdhole/). In his honor I have inserted words for three of his songs, reprinted from https://soils.wisc.edu/wp-content/uploads/2013/11/soil_songs.pdf.

***Where Have All the Bedrocks Gone?***

*Francis D. Hole, 1985 (Tune: Where Have All the Flowers Gone?)*

*Where have all the bedrocks gone?*
*Long time weathering.*
*Where have all the bedrocks gone?*
*That formed so long ago.*
*Where have all the bedrocks gone?*
*Gone to residuum...*
*and to sediments*
*and to the vital soils!*

***You Are My Soil, My Only Soil***

Francis D. Hole, 1985 *(Tune: You Are My Sunshine)*

You are my soil, my only soil;
You keep me vital night and day.
This much I know, friend,
You do support me;
Please don't erode my life's soil away!

***Some Think That Soil Is Dirt***

Francis D. Hole, 1985 *(Tune: Funiculi)*

Some think that. Soil is dirt and quite disgusting
This is not true.
This is not true.
Some think that it makes thee air all brown and dusty
Good dust's in me!
Good dust's in you!

Praise Mother Earth, she is our earthly Mother
She gives us bread.
She gives us bread.
Praise ground, the holy ground that softly under
Our feet that tread.
Our feet that tread.
Vigor, Vigor from the soil does flow;
Roots and life are teeming down below.
No wonder that the land's so green,
The ferns and flowers so fresh and clean!
Soil is everywhere;
From it sweet blessings gently flow.

The "Dig It" exhibit, State Soil designations, and soil science leaders such as Dr. Hole all helped build public awareness and interest in soil health, which has been defined as "the continued capacity of the soil to function as a vital living ecosystem that supports plants, animals, and humans" (USDA-NRCS, 2019). Throughout the second decade of the 21$^{st}$ century, soil health efforts were also enhanced by corporate investments, perhaps driven by economics of sustainability, but oh so important. Through these combined efforts, soil health has evolved from basic principles of soil conservation and stewardship, as well as from prior research and technology transfer efforts including those associated with soil pedology, soil tilth, soil condition, soil management, soil care, soil quality, soil productivity, soil resilience, soil security, and soil degradation. For me, there are two subtle differences between soil health and those efforts. They are: (i) increased emphasis on soil biology because of new tools, techniques, and significant advancements in our understanding of genetics, genomics, and organismal interactions; and (ii) an integration of soil biological, chemical, and physical property and process data and information through holistic assessments.

Recognizing the soil health advancements during the past two decades, these two books, Approaches to Soil Health Analysis (Volume 1) and Laboratory Methods for Soil Health Analysis (Volume 2) were proposed in 2017 to update two SSSA books [Defining Soil Quality for a Sustainable Environment (SSSA Special Publication No. 35) and Methods for Assessing Soil Quality (SSSA Special Publication No. 49)] published during the mid-1990s. Volume 1 is designed to provide an update on how soil health emerged from soil quality during the past two decades. Some of those changes included formation of the Natural Resources Conservation Service (NRCS) Soil Health Division by the USDA, creation of the Soil Health Institute (SHI) by the Noble and Farm Foundations, and numerous private-sector investments beginning with the National Corn Growers Soil Health Partnership (SHP). The second volume is intended to provide current (2020), science-based perspectives and methods for measuring selected soil properties and processes that are emerging as indicators for soil health assessments.

Without question, the soil health concept has been evolving steadily since it was first introduced in the 1970s. Therefore, these volumes are in no way viewed as the final word regarding soil health but rather "works in progress" fully recognizing that improvements and refinements in our knowledge-base and assessment methods are inevitable and continue to be discovered. Editing these Volumes has provided gratification following a professional career that attracted my attention nearly fifty years ago. I recognize any personal contributions to soil health efforts are minuscule considering all who have contributed to these efforts since the time of Plato. None-the-less, it is my hope that as my mentor and friend W. E. Larson (1921-2013) stated many times, each of us needs to do what we can to help humankind understand and appreciate the thin mantle that stands between us and starvation.

Building upon the inspiration of soil science leaders such as Hugh Hammond Bennett, William E. Larson, and John W. Doran, I decided to end these volumes with an Ode to Soil Health. My intent is to honor the living soil with its multitude of functions that influence life on Earth in a multitude of ways. My hope is that by pausing, looking around, and seeing the unique beauty of the "Earth-only" living material known as soil, the importance of this fragile resource will be recognized. Perhaps by enticing those who have absolutely no idea of what lies beneath their feet (clay, sand, or dirt) as well as those who already know and recognize the connection between soil and life with a bit of humor, we will all pause, look more closely at what we walk on each day and truly strive to enhance soil health by mitigating soil erosion, nutrient depletion, loss of soil organic matter, compaction, contamination, or any other peril threatening these living, life-sustaining, fragile resources.

## Notes

1 Francis Doan Hole (1913-2002) was a soil science professor at the University of Wisconsin when I was an undergraduate. As an American pedologist, educator, and musician, his love for our fragile soil resources was documented by mapping the extent of soils and their properties throughout Wisconsin and using inventive lectures and musical performances to communicate and popularize the field of soil science. He is recognized as the leader for establishing the Antigo silt loam as the official state soil of Wisconsin in 1983 when the legislature passed Wisconsin Act 33. Hole and fellow advocates argued for many years that soils are a natural resource that took thousands of years to form after glaciers retreated from Wisconsin between 12,500 and 15,000 YBP. Dr. Hole relentlessly advocated that soil was essential to Wisconsin's economy and the very foundation of terrestrial life for residents of the

State and elsewhere. Within the Soil Science Society of America and around the world, Francis is remembered as an "Ambassador of Soils" and "Poet Laureate of Soil Science". His life is indeed a tribute to the goals and aspirations of Soil Health.

## References

Torres, A., J. Liu, J. Brandt, and K. Lear. 2017. The World is Running Out of Sand. Smithsonian Magazine (The Conversation). https://www.smithsonianmag.com/science-nature/world-facing-global-sand-crisis-180964815/

USDA-Natural Resources Conservation Service (NRCS). 2019. The Basics of Addressing Resource Concerns with Conservation Practices within Integrated Soil Health Management Systems on Cropland, by D. Chessman, B.N. Moebius-Clune, B.R. Smith, and B. Fisher. Soil Health Technical Note No. 450-04. Available on NRCS Electronic Directive System. Washington, DC. https://directives.sc.egov.usda.gov.

Wikipedia. 2020. Francis D. Hole. https://en.wikipedia.org/wiki/Francis_D._Hole

Yang, J. E., M. B. Kirkham, R. Lal, and S. Huber (eds.). 2018. Global Soil Proverbs: Cultural Language of the Soil. Catena-Schweizerbart, Stuttgart, Germany.

**An Ode to Soil Health**

Some may ask, why dirt is important to us,
Isn't soil something about which only a scientist would fuss?

Global wealth is derived from gold, silver, and oil.
Other than farmers, who cares about soil?

On fingers, windows, shirts, or in piles blocking our way,
Is soil anything more than sand and clay?

The answer is yes, for when an ancient scroll we do unroll,
From "dust we came and to dust we'll go," many extoll.

*Adama* and *Hava* refer to man and wife,
Reflect soil health, when translated to "Soil and Life."

The thin mantle in which humans for food do toil,
Is, what Earth Scientists call soil.

Sand, silt and clay particles, classified by size,
Bound by chemistry, physics, and biology before our eyes.

People think sand is found only on the beach,
For .05 to 2 mm particles, a morphologist does reach.

Silt at 2 to 50 microns in size, often moves by air,
Creating loess hills or getting caught in your hair.

Clay at less than 2 microns in size,
Its 1:1 or 2:1 mineralogy controls more than you realize.

Soil particles derived from rocks,
Link together as aggregate building blocks.

Bound by biological glues and chemical forces,
Aggregate soil structure has the strength of horses.

From the soil surface to underlying bedrock,
Pedons reflect all horizons in stock.

Granular, prismatic, and massive soil structure,
Soil Scientists view this, as intricate sculpture.

Water infiltration, air exchange, and runoff,
Enhancing soil health cannot be put off.

Acidity, salinity, and soil solution transport,
Productivity, this chemistry does support.

Soil chemistry and physics are fun to explore,
But dynamic soil biology adds much more.

In every gram of soil, many organisms are living,
Some cause problems, but others are life-giving.

In 1928 Fleming introduced penicillin to the scene,
Soil metagenomics revealed antibacterial violacein.

A symbiotic soil *Pseudomonad,* has for you,
Produced Pederi, an anticancer drug too.

Multitudes of bacteria and innumerable fungi that we can't see,
A myriad of algae and numberless nematodes also live free.

An army of earthworms and countless protozoa too,
Mychorrhizae and actinomycetes working just for you.

Millions of micro- and macro-fauna too,
Ants, termites, millipedes, earthworms, slugs, and snails to name a few.

Responding to temperature and drought,
Changes in soil carbon, these organisms bring about.

They protect food, feed and fiber from abiotic stress,
Without pathogen resistance, earth would be a mess.

Soil microbes decompose our waste,
Creating chitin, glomalin and other compounds in its place.

Glucosidase, PLFA and EL-FAME,
Soil Health indicators have several names.

Through tillage, irrigation, and fertilizing,
Microbes are also affected, by livestock grazing.

Carbon sequestration and humification,
Without those processes, GHG would put earth on permanent vacation.

Buffering a continuously changing climate,
Dynamic, living soil keeps earth in alignment.

A glimpse of living soil, I have given you,
But wait there is even more, these fragile resources do.

Soil supports plant roots and exchange of water and air,
Cycles nutrients, filters, buffers and sustains biological diversity with flare.

It produces food, feed, fiber and fuel when given good care,
But erodes and releases GHGs, if sustainable practices are not there.

To make living soil less stealth,
These Volumes were written to promote Soil Health.

To the end of this ode, we have now come,
Hopefully your quest for Soil Health has just begun.

Printed in the USA/Agawam, MA
September 16, 2026

910661.220